Letts

AS

VISUAL
REVISION
GUIDE

SUCCESS

PHYSICS

Brian Arnold

Contents

Forces and motion 1

Forces and motion 2

Waves

Electricity

Quantum physics

Kinetic Theory

The nuclear atom and radioactivity

Scalars and vectors

Scalars

Scalar quantities have <u>magnitude</u> but no direction.
EXAMPLES
mass, volume, energy, power, charge

To add scalars, simply add the quantities together.
EXAMPLES
$3\,kg + 2\,kg = 5\,kg$
$10\,cm^3 + 40\,cm^3 = 50\,cm^3$

EXAMINER'S TOP TIP
Remember to include units with all your answers.

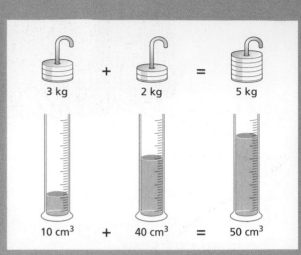

3 kg + 2 kg = 5 kg

10 cm³ + 40 cm³ = 50 cm³

Vectors

Vector quantities have <u>magnitude</u> **and** <u>direction.</u>

EXAMPLES

 displacement

acceleration

force

momentum

When adding vectors, the direction of the quantity as well as its magnitude must be taken into account.

EXAMPLE
displacement from A is 5 m   displacement from B is 4 m   total displacement from A is 9 m

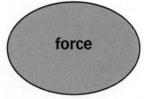

$$A \xrightarrow{\ 5\,m\ } B + B \xrightarrow{\ 4\,m\ } C = A \xrightarrow{\qquad 9\,m \qquad} C$$

but
displacement from A is 5 m   displacement from B is –4 m   total displacement from A is (5 – 4) m, i.e. 1 m

$$A \xrightarrow{\ 5\,m\ } B + C \xleftarrow{\ 4\,m\ } B = A \xrightarrow{\ 1\,m\ } C$$

Note Weight is a force and therefore a vector.

Adding vectors

Adding vectors that are not in line using scale diagrams

Method A

1. Draw first vector to scale and in the correct direction.

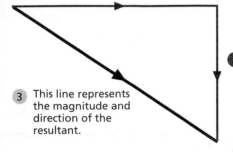

2. Draw second vector at the end of the first. Draw it to scale and in the correct direction.

3. This line represents the magnitude and direction of the resultant.

Method B

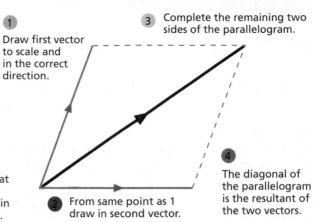

1. Draw first vector to scale and in the correct direction.

2. From same point as 1 draw in second vector.

3. Complete the remaining two sides of the parallelogram.

4. The diagonal of the parallelogram is the resultant of the two vectors.

EXAMPLE

A force of 10 N is applied to an object O due north.
A second force of 10 N, in a direction 10° north of east, is applied to O. What is the resultant force experienced by O?

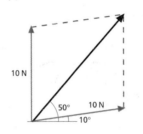

Resultant force is 15.3 N in a direction 50° north of east.

EXAMPLE

A woman walks 8.0 m east and then 6.0 m south. What is her displacement from her starting point?

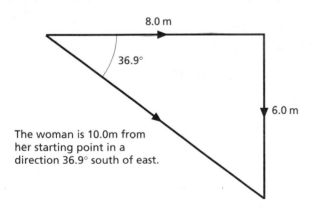

The woman is 10.0m from her starting point in a direction 36.9° south of east.

Adding vectors by calculation

If the vectors are at right angles, their resultant can be found using Pythagoras' Theorem.
For the example with the woman walker:

$$R^2 = 8.0^2 + 6.0^2$$
$$R^2 = 100$$
$$\therefore R = 10 \text{ m}$$
$$\tan \theta = \frac{6.0}{8.0} = 0.75$$
$$\therefore \theta = 36.9°$$

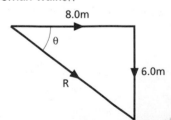

Quick test

1. What is a scalar quantity? Give one example.

2. What is a vector quantity? Give one example.

3. A man on board a ship runs northwards at 5.0 m s⁻¹ while the ship is moving westwards at 7.0 m s⁻¹. Determine the man's resultant velocity.

4. If the ship in question 3 changes direction so that it is travelling southwards, what is the man's resultant velocity assuming his velocity and the ship's speed remain unchanged?

1. a quantity that has only magnitude, for example speed, volume, etc. 2. a quantity that has magnitude and direction, for example velocity, force, etc. 3. 8.6 m s⁻¹ 35.5° north of west 4. 2.0 m s⁻¹ south

5

Resolving vectors

● Sometimes it is useful to be able to split a single vector into two <u>components</u>. These will often be the horizontal and vertical components. This can be done by drawing scale diagrams or by using a calculator.

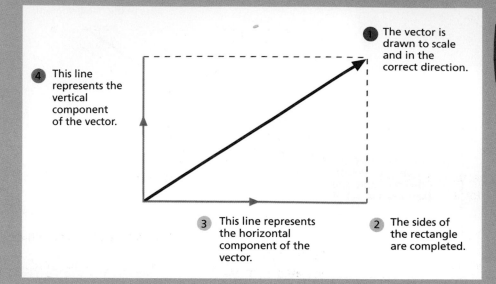

④ This line represents the vertical component of the vector.

① The vector is drawn to scale and in the correct direction.

③ This line represents the horizontal component of the vector.

② The sides of the rectangle are completed.

Resolving using a scale diagram

EXAMPLE

This aircraft is travelling at 300 km h⁻¹ and at an angle of 40° to the horizontal. What are the horizontal and vertical velocities of the aircraft?

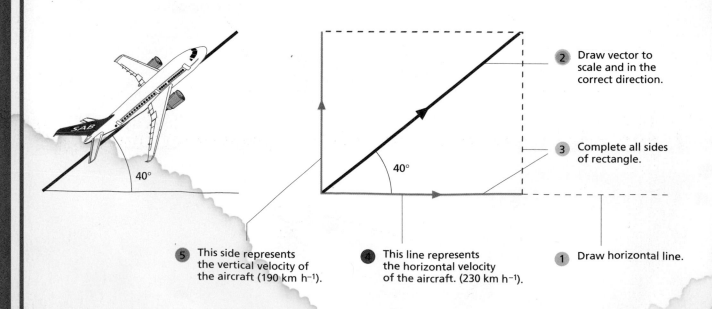

② Draw vector to scale and in the correct direction.

③ Complete all sides of rectangle.

① Draw horizontal line.

⑤ This side represents the vertical velocity of the aircraft (190 km h⁻¹).

④ This line represents the horizontal velocity of the aircraft. (230 km h⁻¹).

Resolving vectors using calculation

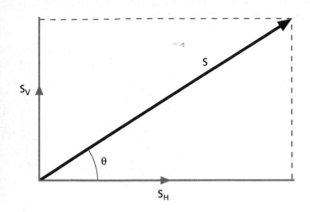

The horizontal component of the vector is $S_H = S \cos \theta$
The vertical component of the vector is $S_V = S \sin \theta$

EXAMPLE

For the aircraft in the previous example:
The horizontal velocity $S_H = 300 \cos 40 = 230$ km h^{-1}
The vertical velocity $S_V = 300 \sin 40 = 190$ km h^{-1}

EXAMPLE

The diagram shows an object weighing 40 N suspended on two strings. Calculate the tension of each of the strings.

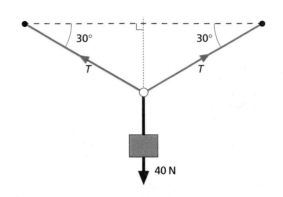

Because the object is not falling, the vertical forces must be balanced.
Resolving vertically:

$$T \cos 60 + T \cos 60 = 40 \, N$$
$$2T \cos 60 = 40 \, N$$
$$T = 40 \, N$$

Quick test

1 Resolve these vectors into their horizontal and vertical components using a scale diagram.

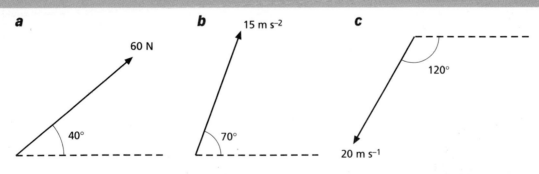

2 Resolve these vectors into their horizontal and vertical components using calculation.

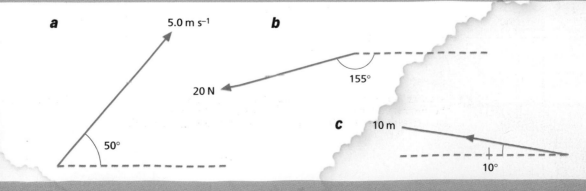

1. a) 46 N, 39 N   b) 5.1 m s^{-2}, 14 m s^{-2} c) −10 m s^{-1}, −17 m s^{-1}
2. a) 3.2 m s^{-1}, 3.8 m s^{-1} b) −18 N, −8.5 N   c) −9.8 m, 1.7 m

7

Describing motion using graphs

Displacement-time graphs

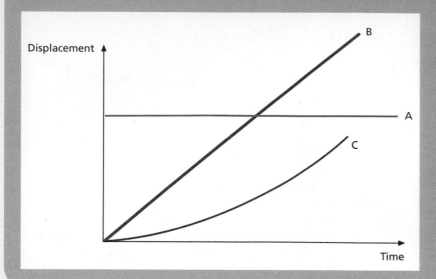

A Object is stationary. Gradient of graph is zero.

B Object is moving at a constant velocity. This velocity is equal to the gradient of the graph.

C Object is accelerating. The gradient of the graph is increasing.

Velocity-time graphs

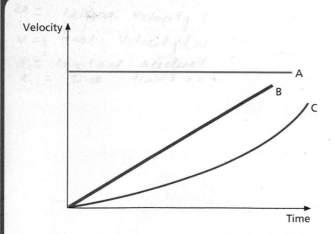

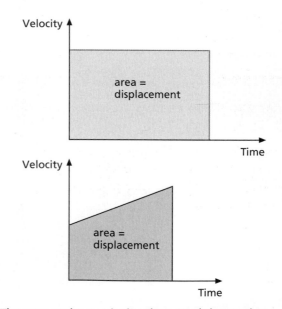

A Object is moving at a constant velocity. Gradient of graph is zero.

B Object has constant acceleration. The acceleration is equal to the gradient of the graph.

C Object's acceleration is increasing. The gradient of the graph is changing.

The area under a velocity–time graph is equal to displacement.

Acceleration-time graphs

A Acceleration of object is increasing uniformly with time.
B Acceleration of object is decreasing non-uniformly with time.

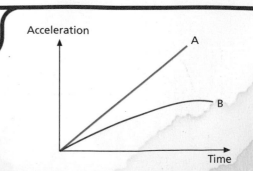

Further examples

An object falling freely on the Earth that is experiencing no resistance or frictional forces

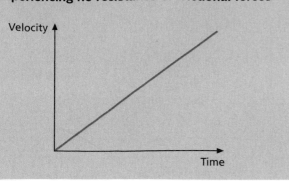

An object falling through air (or a liquid)

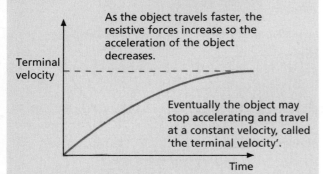

As the object travels faster, the resistive forces increase so the acceleration of the object decreases.

Terminal velocity

Eventually the object may stop accelerating and travel at a constant velocity, called 'the terminal velocity'.

An object that falls and then bounces on a hard floor

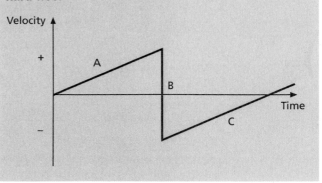

A Ball bearing has constant acceleration.
B Ball collides with floor and velocity decreases and changes direction almost instantly.
C Ball rebounds from floor with a velocity almost equal in magnitude to that at collision but in the opposite direction, i.e. it is a negative velocity. As the ball rises it decelerates at the same rate as it accelerated during its fall. Line A and line C therefore have the same gradient.

EXAMINER'S TOP TIP
Don't press too hard when drawing a graph. Then, if you don't like what you have drawn and you have to do it again you can rub the line out without it looking a mess. Think ... sharp pencils and thin lines.

HINT Before answering any questions about graphs, look carefully at the axes. It is easy to misread a displacement–time graph for a velocity–time graph and vice versa.

Quick test

1 Sketch a velocity–time graph for a free-fall parachutist from the time she jumps from the aircraft to the time she lands on the ground.

2 Sketch a velocity–time graph for a pendulum.
HINT At time $t=0$, the pendulum has been pulled to one side and is about to be released.

3

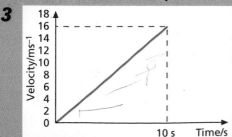

a Calculate the acceleration of the object described in the graph above.

b Calculate the distance travelled by the object in 10 s.

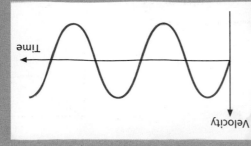

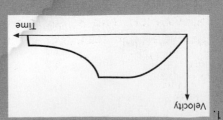

2.

3. a) 1.6 m s^{-2} b) 80 m

1.

Equations of motion

Speed and acceleration

By definition

$$\text{speed} = \frac{\text{change in distance}}{\text{time taken}}$$

$$\text{velocity} = \frac{\text{change in displacement}}{\text{time taken}} \quad \text{or} \quad v = \frac{\Delta s}{\Delta t}$$

$$\text{acceleration} = \frac{\text{change in velocity}}{\text{time taken}} \quad \text{or} \quad a = \frac{\Delta v}{\Delta t} \quad \text{or} \quad a = \frac{v-u}{t}$$

where a is the acceleration, v the final velocity and u the initial velocity.

Further equations

There are several other equations that can be used to describe and analyse the movement of an object provided that its **acceleration is constant**. These are:

$s = $ displacement (m)
$u = $ initial velocity (ms⁻¹)
$v = $ final velocity (ms⁻¹)
$a = $ constant acceleration (ms⁻²)
$t = $ time interval (s)

$$v = u + at$$
$$s = ut + \tfrac{1}{2}at^2 \quad \text{and} \quad s = \tfrac{1}{2}(u+v)t$$
$$v^2 = u^2 + 2as$$

The derivation of these equations may not be needed but you must be able to use them.

EXAMPLE

If a car with an initial velocity of 5.0 m s⁻¹ accelerates at 4.0 m s⁻² for 10 s, calculate its final velocity.

$v = ?$
$u = 5.0$ m s⁻¹
$a = 4.0$ m s⁻²
$t = 10$ s
s is not needed

Using $v = u + at$

$$v = 5.0 + 4.0 \times 10$$

$$v = 45 \text{ m s}^{-1}$$

EXAMPLE

A motorcyclist starting from rest accelerates at 1.5 m s⁻² for 8.0 s. Calculate how far he travels during this time.

v is not needed
$u = 0$ m s⁻¹
$a = 1.5$ m s⁻²
$t = 8.0$ s
$s = ?$

Using $s = ut + \tfrac{1}{2}at^2$

$$s = 0 + \tfrac{1}{2} \times 1.5 \times 8.0^2$$

$$s = 48 \text{ m}$$

EXAMPLE

An aircraft has a take-off speed of 80 m s⁻¹. If the runway is 1.5 km long, what is the minimum acceleration the aircraft must have in order to take off?

$v = 80$ m s⁻¹
$u = 0$ m s⁻¹
$a = ?$
t is not needed
$s = 1.5 \times 1000$ m

Using $v^2 = u^2 + 2as$

$$a = \frac{v^2 - u^2}{2s}$$

$$a = \frac{80^2 - 0^2}{2 \times 1.5 \times 1000}$$

$$a = 2.1 \text{ m s}^{-2}$$

EXAMINER'S TOP TIP
When tackling a question:
• read the problem carefully
• identify the quantities given
• identify the quantity you need to find
• now select the equation that contains all these quantities.

Measuring the acceleration due to gravity

If the time it takes for a released object to fall a distance (s) is known, the acceleration due to gravity (g) can be found using $s = ut + \frac{1}{2}at^2$.

If the initial velocity of the ball bearing is 0,

then $s = \frac{1}{2}gt^2$

or

$$g = \frac{2s}{t^2}$$

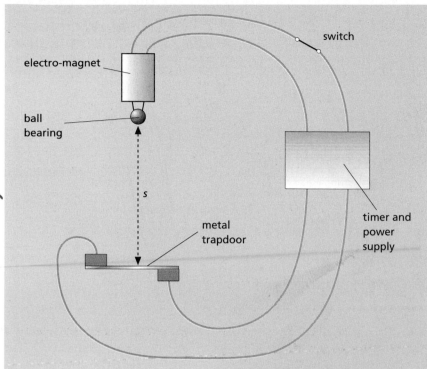

- electro-magnet
- ball bearing
- switch
- timer and power supply
- metal trapdoor
- s

- When the switch is opened, the electromagnet is turned off.

- The ball bearing is released and the timer started.

- After falling a distance s, the ball bearing knocks open a metal trapdoor that stops the timer.

EXAMINER'S TOP TIP
Double-check that you have given the correct units with your answers. Lots of method marks are available in this topic, even with the wrong answer, so show all your working!

- The experiment should be carried out several times at different heights and an average for g found.
- Using a small dense object such as a ball bearing means that frictional forces due to air are negligible.
- For greater accuracy, s should be as large as possible.

Quick test

1 Calculate the final speed of a ball bearing that falls from rest for 5.0 s. ($g = 9.8\,\text{m s}^{-2}$)

2 A rocket travelling at $10\,\text{m s}^{-1}$ accelerates at $15\,\text{m s}^{-2}$ for 10 s. Calculate the final speed of the rocket.

3 A sprinter at the beginning of a race accelerates at $2.0\,\text{m s}^{-2}$ for the first 25 m. Calculate the speed of the sprinter after running 25 m.

4 A ball is thrown vertically upwards with a velocity of $30\,\text{m s}^{-1}$. How long will it be before the ball returns to its starting point? ($g = 9.8\,\text{m s}^{-2}$)

5 A car travelling at $30\,\text{m s}^{-1}$ accelerates at $1.0\,\text{m s}^{-2}$ for 20 s. Calculate

  a the final speed of the car

  b the distance travelled during this 20 s

1. $49\,\text{m s}^{-1}$ 2. $160\,\text{m s}^{-1}$ 3. $10\,\text{m s}^{-1}$ 4. 6.1 s 5. a $50\,\text{m s}^{-1}$ b 800 m

Projectiles

- If we throw a ball horizontally from the top of a tower or cliff, we see that its motion is affected by gravity. As the ball moves away from us it begins to move vertically, i.e. it falls.
- If a second ball is dropped at the instant the first is thrown, their motions can be compared.

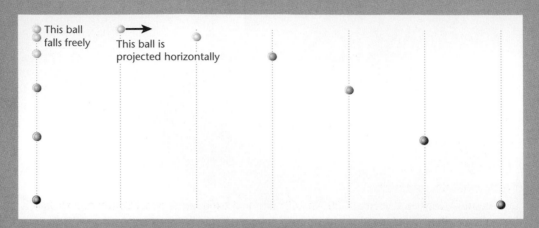

- This ball falls freely
- This ball is projected horizontally

- The first ball travels away from us at a constant horizontal velocity.
- The vertical motion of both balls is identical. Both have an acceleration of 9.8 m s⁻² (assuming air resistance is negligible).

We can see from the above that the horizontal motion and the vertical motion of a projectile are independent of each other. This is important because it means we can break down the motion of a projectile into its horizontal and vertical components and then analyse these separately.

Projectile problems

EXAMPLE 1

A ball is thrown with a horizontal velocity of 10 m s⁻¹ from a tower 200 m above the ground. How far from the base of the tower will the ball hit the ground?

To solve the problem, we must first calculate the time that the ball travels before hitting the ground. We can do this by considering the vertical motion of the ball.

Consider now the horizontal motion of the ball.

Whilst the ball has been falling it has also been moving horizontally at a constant speed of 10 m s⁻¹.

The total horizontal distance travelled (s) is therefore

$s = ut = 10 \times 6.4 = 64\,\text{m}$

Using $s = ut + \frac{1}{2}at^2$

$s = 200\,\text{m},$
$u = 0,$
$t = ?$
$a = 9.8\,\text{m s}^{-2}$

therefore $200 = 0t + \frac{1}{2} \times 9.8 \times t^2$

therefore $t^2 = \dfrac{400}{9.8}$

therefore $t = 6.4\text{s}$

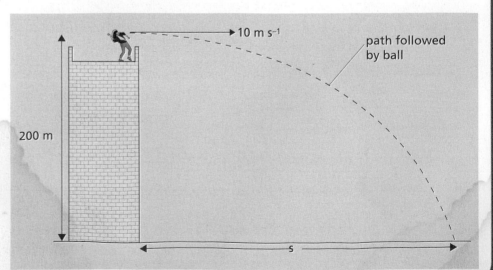

10 m s⁻¹

path followed by ball

200 m

s

Projectile problems

EXAMPLE 2

A shell is fired from a gun at an angle of 60° to the horizontal and with a speed of 200 m s⁻¹. Calculate.
a) the total time the shell travels before striking the ground
b) the range (horizontal distance travelled) of the shell.

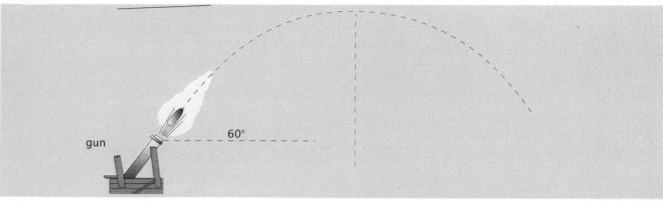

a) Consider the vertical motion of the shell.

The shell will have an initial vertical velocity of 200 sin 60° and will experience acceleration due to gravity of –9.8 m s⁻². From these two pieces of information, we can calculate the time it takes for the shell to reach its maximum height, i.e. when the vertical velocity of the shell is zero.

Using	$v = u + at$

$$0 = 200 \sin 60° - 9.8t$$

therefore $\quad t = \dfrac{200 \sin 60°}{9.8}$

$$t = 17.66 \approx 17.7 \text{ s}$$

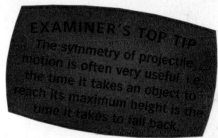

EXAMINER'S TOP TIP
The symmetry of projectile motion is often very useful, i.e. the time it takes an object to reach its maximum height is the time it takes to fall back.

As the diagram above shows, the motion of the shell is symmetrical, i.e. the time it takes to return to the ground is also 17.7 s. Therefore, the total time the shell is in the air is 35s (or 35.4s).

b) Consider now the horizontal motion of the shell.

The shell is travelling at a constant horizontal speed of 200 cos 60° = 100 m s⁻¹

The total horizontal distance travelled by the shell (s) = ut = 100 × 35 = 3500 m

EXAMINER'S TOP TIP
Units again .... don't forget them!

Quick test

1 Calculate the horizontal velocity for each of these.

  a A ball thrown with a speed of 20 m s⁻¹ and at an angle of 50° to the horizontal.

  b An arrow shot with a speed of 50 m s⁻¹ at an angle of 20° to the vertical.

2 Calculate the vertical velocity for each of these.

  a An aircraft travelling with a speed of 60 m s⁻¹ and at an angle of 30° to the horizontal.

  b A shell fired with a speed of 150 m s⁻¹ and at an angle of 55° to the vertical.

3 A ball is thrown upwards at an angle of 80° to the horizontal and with an initial velocity of 40 m s⁻¹. Ignoring air resistance, calculate the time the ball is in the air. (g = 9.8 m s⁻²)

4 Find the range of a shell fired at an angle of 40° to the vertical and with a speed of 100 m s⁻¹. (Assume that air resistance is negligible.)

5 An aircraft climbs vertically at a velocity of 50 m s⁻¹ and during the same period travels a horizontal distance of 1500 m in 7.5 s. Combine these two vectors to calculate the velocity of the aircraft.

1. a) 13m s⁻¹ b) 17m s⁻¹ 2. a) 30m s⁻¹ b) 86m s⁻¹ 3. 8.0 s 4. 1000m 5. 206 m s⁻¹ 14° above the horizontal

Newton's laws of motion

Newton's first law

If there is no resultant force acting upon an object, that object will:
remain at rest *or*

if it is moving, it will continue to move with a constant velocity, i.e. at a constant speed and in a straight line.

EXAMPLES

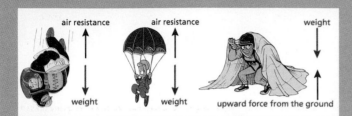

When the frictional forces equal the weight, the parachutist will fall at a constant velocity called his **terminal velocity**.

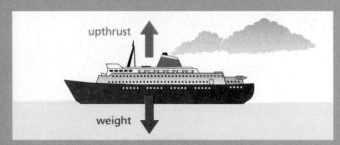

The **upthrust** from the water and the weight are equal and opposite. There is no resultant vertical force so the ship does not sink.

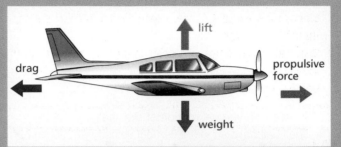

The propulsive force and the drag force acting on this aircraft are balanced.
The lift force and the weight are also balanced.
The aircraft therefore travels at a constant horizontal velocity and at the same height.

EXAMINER'S TOP TIP
Remember ... No motion, no resultant force. Constant velocity ... no resultant force.

F = ma

When a body is acted upon by an external force it accelerates. The size of the acceleration is proportional to the resultant force.

This can be described by the equation
$$F = ma$$

F = resultant force applied to object in N
m = mass of object in kg
a = acceleration of object in $m\,s^{-2}$

EXAMINER'S TOP TIP
Try to think of examples that illustrate $F = ma$, for example accelerating rockets, crashing cars, etc.

EXAMPLE
A resultant force of 1000 N is applied to a car of mass 500 kg.
Calculate the acceleration of the car.

$$F = ma$$
$$a = \frac{F}{m}$$
$$a = \frac{1000}{500}$$
$$a = 2.0\ m\,s^{-2}$$

Surviving a collision

- The forces due to deceleration exerted during a crash can cause serious injuries or be fatal.
- From $F = ma$

$$F = m\frac{v - u}{t}$$

we can see that the longer the period of deceleration, t, the smaller the deceleration force exerted on the car and its passengers.
- Modern cars have crumple zones. During a collision these zones become squashed, so that deceleration takes place over a longer time and the forces exerted on people in the car are therefore smaller.
- Seat belts have some elasticity. In a crash, they will stretch a little, which increases the time the belt applies a force to the person. As a result, the size of the force is less.

crumple zone   crumple zone

Newton's third law

To every action there is an equal and opposite reaction.

EXAMPLES

Similar poles repel. The forces exerted on the magnets are equal in magnitude but in opposite directions.

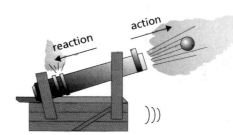

reaction   action

When a cannon ball is fired, a force is applied to it. An equal and opposite force (recoil) is applied to the cannon.

Quick test

1 Compare the horizontal forces applied to a car travelling at its top speed along a straight, flat road.

2 Calculate the propulsive force on a firework rocket that has a mass of 0.1 kg and an initial acceleration of 20 m s⁻².

3 A sprinter of mass 80 kg running at 10 m s⁻¹ slows to 2 m s⁻¹ in 4.0 s. Calculate the braking force he exerts on himself.

4 Explain, using Newton's third law, what will happen if a man steps off an unsecured rowing boat on to a bank.

4. As he pushes himself forward towards the bank, the boat will be pushed away from the bank (action and reaction).
3. 160 N
2. 2.0 N
1. The propulsive force and resistive forces are balanced, i.e. there is no resultant force.

Momentum

- The <u>momentum</u> of an object (p) is defined as being the product of its mass (m) and its velocity (v):

$$p = m \times v$$

Momentum is a vector. It has size and direction and is measured in kg ms⁻¹.

EXAMPLE

Calculate the momentum of an object which has a mass of 10 kg and is moving due north at a speed of 20 m s⁻¹.

$p = m \times v = 10 \times 20 = 200$ kg m s⁻¹ due north.

Law of motion and momentum

Newton's second law of motion and momentum

This states that the rate of change of momentum of a body is directly proportional to the resultant force applied to it and is in the direction of the force, i.e.

$$F \propto \frac{\Delta p}{\Delta t}$$

If these quantities are measured in SI units we can write

$$F = \frac{\Delta p}{\Delta t} \qquad \text{or} \qquad F = \frac{mv - mu}{t} \qquad \text{or} \qquad F = ma$$

EXAMINER'S TOP TIP
The link between momentum and Newton's second law of motion is important.

EXAMPLE

A train of mass 1.0×10^5 kg slows from a velocity of 100 m s⁻¹ to 10 m s⁻¹ in 3 minutes. Calculate

a) the change in the train's momentum

Change in momentum $= mv - mu$
$= 1.0 \times 10^5 \times 10 - 1.0 \times 10^5 \times 100$
$= -9.0 \times 10^6$ kg m s⁻¹

b) the average braking force applied to the train

$$F = \frac{mv - mu}{t}$$

$$F = \frac{9.0 \times 10^6}{180}$$

$$F = 5.0 \times 10^4 \text{ N}$$

Impulse

If we rearrange Newton's second law

$$F = \frac{mv - mu}{t}$$

we obtain the equation

$$Ft = mv - mu$$

The product Ft is known as the **impulse of a force** and is equal to the change in momentum it causes.

Law of conservation of momentum

- When two or more objects collide they exert forces on each other.
- These forces cause the momentum of each of the objects to change.
- The law of conservation of momentum describes how the momenta change. It states that

the total momentum of the objects before the collision must be equal to the total momentum after the collision, as long as there are no external forces.

$$m_1u_1 + m_2u_2 = m_1v_1 + m_2v_2$$

EXAMPLE

Calculate the final velocity of the 5.0 kg ball after the collision described below.

before

10.0 m s⁻¹   5.0 m s⁻¹

5.0 kg   1.0 kg

after

V_1 10.0 m s⁻¹

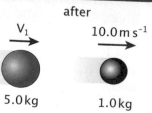

5.0 kg   1.0 kg

Using $\quad m_1u_1 + m_2u_2 = m_1v_1 + m_2v_2$

$5.0 \times 10.0 + 1.0 \times 5.0 = 5.0 \times v_1 + 1.0 \times 10.0$

$50.0 + 5.0 = 5.0 \times v_1 + 10.0$

$$v_1 = \frac{45.0}{5.0} = 9.0 \text{ m s}^{-1}$$

The law can also be applied to objects that are moving in opposite directions before or after a collision.

Further examples

EXAMPLE

Calculate the final velocity of the 1.0 kg ball after the collision described below.

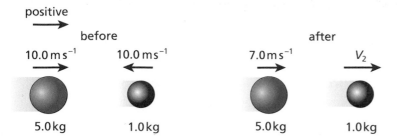

positive

before

after

10.0 m s⁻¹ → 10.0 m s⁻¹ ← 7.0 m s⁻¹ → V_2

5.0 kg 1.0 kg 5.0 kg 1.0 kg

Using

$$m_1u_1 + m_2u_2 = m_1v_1 + m_2v_2$$
$$5.0 \times 10.0 + 1.0 \times -10.0 = 5.0 \times 7.0 + 1.0 \times v_2$$
$$50.0 - 10.0 = 35.0 + 1.0\,v_2$$
$$v_2 = \frac{50.0 - 10.0 - 35.0}{1.0} = 5.0 \text{ m s}^{-1}$$

EXAMINER'S TOP TIP

Remember that, when objects collide, they may have positive or negative velocities before or after the collision.

In some situations the momentum of two objects before they interact may be zero. For example, before this cannon is fired the total momentum of the cannon and the cannon ball is zero. After the ball has been fired, the total momentum of the two must still be zero.

EXAMPLE

Calculate the recoil velocity of the cannon.

Using
$$m_1u_1 + m_2u_2 = m_1v_1 + m_2v_2$$
$$0 = 30 \times 1.0 + 100 \times v_2$$
$$v_2 = \frac{-30}{100}$$
$$v_2 = -0.30 \text{ m s}^{-1}$$

30 m s⁻¹

1.0 kg

V_2

100 kg

Elastic and inelastic collisions

During a collision the kinetic energies of the interacting bodies may change.

If the sum of the kinetic energies of the bodies before a collision is equal to the sum of their kinetic energies after the collision this is described as an **elastic collision**.

In an **inelastic collision** the kinetic energies of the colliding bodies are *not* conserved.

Quick test

1 A railway wagon with a mass of 2000 kg travelling at a speed of 25 m s⁻¹ collides with a second wagon with a mass of 1000 kg travelling at a speed of 10 m s⁻¹ in the same direction. After the collision the wagon with the smaller mass continues in its original direction at a speed of 30 m s⁻¹. Calculate the new speed of the 2000 kg wagon.

2 A railway wagon with a mass of 1000 kg travelling at a speed of 20.0 m s⁻¹ collides with a second, identical, wagon which is stationary. After the collision the two wagons stick together. Calculate the new speed of the wagons.

3 Calculate the change in momentum of a ball when a cricket bat applies a force of 60 N to it for a time of 0.25 s.

4 Explain why the recoil velocity of a cannon is much smaller than the firing velocity of a cannon ball.

1. 15 m s⁻¹ 2. 10 m s⁻¹ 3. 15 kg m s⁻¹ 4. The mass of the cannon is much larger than the mass of the ball and so, to conserve momentum, the recoil velocity of the cannon must be much smaller than the firing velocity of the cannon ball.

19

Moments and equilibrium

- **The turning effect of a force is called a <u>moment</u>.**
  The moment of a force can be calculated using this equation:

  moment of force = force × perpendicular distance from pivot

  EXAMPLES

  Moment = $F \times d$

  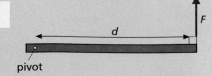

  HINT As an alternative method
  you could use the
  component of *F* that is
  perpendicular to *d*.
  i.e. $F \cos \theta$
  ∴ Moment = $F \cos \theta \times d$
  = $Fd \cos \theta$

  Moment = $F \times d \cos \theta$
  = $Fd \cos \theta$

  EXAMPLE

  Calculate the moment created
  by a force of 50 N applied to a
  crowbar 1.5 m long. The
  direction of the force is at an
  angle of 75° to the crowbar.

  Moment = $F \times d \cos \theta$
  = $50 \times 1.5 \times \cos 15$
  = 72.4 N m

  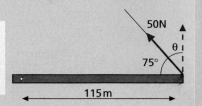

Principle of moments

EXAMINER'S TOP TIP
When calculating the size of a moment, remember it must be the perpendicular distance of the force from the pivot that you use in your equation.

If a body is in equilibrium, the sum of the clockwise moments about any axis (pivot)
is equal to the sum of the anticlockwise moments.

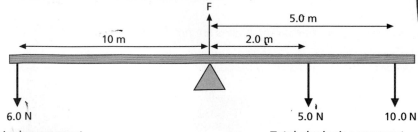

Total anticlockwise moment
= 6.0×10
= 60 N m

Total clockwise moment
= $5.0 \times 2.0 + 10.0 \times 5.0$
= 60 N m

Couples and torques

A **<u>couple</u>** consists of a pair of equal, but opposite, forces that do not act along the
same line, i.e. there is a turning effect but no resultant force.

The **<u>torque</u>** of a couple is the total moment created by the forces.
   Torque = magnitude of one force × perpendicular distance between them
In the example, the torque exerted on the key is $F \times d$.

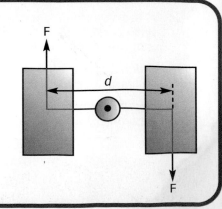

Conditions for equilibrium

For a body to be in equilibrium:
- the sum of the forces acting upon it in any direction must be zero,
- the sum of the moments acting about any point must be zero.

EXAMPLE

A plank of weight 50N resting on two supports

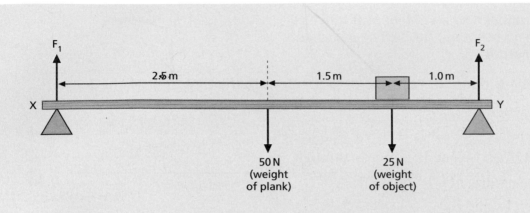

The plank is in equilibrium
sum clockwise moments about X = sum anticlockwise moments about X.

$(50 \times 2.5) + (25 \times 4.0) = F_2 \times 5.0$

$\qquad 225 = F_2 \times 5.0$

$\qquad F_2 = 45 \text{ N}$

The plank is in equilibrium, the total force upwards = total force down.

$F_1 + F_2 = 50 + 25$

$F_1 + 45 = 75$

$\therefore F_1 = 30 \text{ N}$

EXAMINER'S TOP TIP
To check for equilibrium, it is often useful to confirm that the sum of the vertical components of all the forces is zero and that the sum of the horizontal components of all forces is also zero.

Quick test

1 What is a moment?

2 What is a couple?

3 What is a torque?

4 State the conditions necessary for an object to be in equilibrium.

5 Calculate the moment created by the force F shown in the diagram below about the pivot.

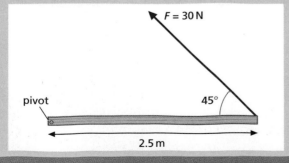

6 Calculate the torque created by the forces shown in the diagram below.

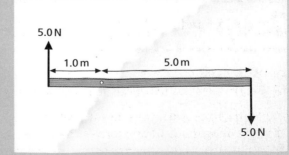

1. the turning effect of a force   2. a pair of equal and opposite forces not acting in line   3. the total moment created by the forces in a couple   4. sum of all forces acting on it in any direction is zero and sum of the moments acting about any point is zero   5. 53 N m   6. 30 N m

21

Work, energy and power

If we apply a force to an object, it may cause the object to move or to change shape. In each of these cases energy is being <u>transferred</u> to the object. The amount of energy transferred is equal to the <u>work done</u> by the force.

If a force F is applied to an object as it moves a distance s, the work done (W) can be calculated using this equation:

> Work done = Force × displacement in the direction of the force
> $$W = Fs$$

Work done is measured in <u>joules (J)</u>. If a force of 1 N is applied to an object as it moves through a distance of 1 m, the work done by the force is 1 J.

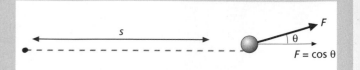

If the force and the displacement of the object are not in the same direction, then the component of the force in the direction of the displacement must be found.

$$W = F \cos \theta \times s$$
$$W = Fs \cos \theta$$

EXAMPLE

A man pulling a boat along the side of a stream applies a force of 100 N in a direction that is at an angle of 20° to the direction in which the boat is moving. Calculate the work done by the man if he moves the boat a distance of 50 m.

$$W = Fs \cos \theta$$
$$W = 100 \times 50 \cos 20°$$
$$W = 4698.4 \text{ J } (4.7 \text{ kJ 2 s.f.})$$

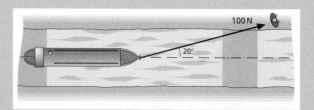

Kinetic energy

An object which is moving has kinetic energy. The kinetic energy it possesses, E_k, can be calculated using this formula.

> $E_k = \frac{1}{2}mv^2$ m is mass of object in kg
> v is speed of object in m s⁻¹

EXAMPLE

Calculate the kinetic energy of a cannon ball of mass 2.0 kg moving at a speed of 40 m s⁻¹.

Using $E_k = \frac{1}{2}mv^2$
$$E_k = \frac{1}{2} \times 2.0 \times 40^2$$
$$E_k = 1600 \text{ J}$$

Potential energy

An object which is above the ground has **gravitational potential energy** E_p because of its position. If its height above the ground changes, then so does its gravitational potential energy. This change in the gravitational potential energy of the object is described by this equation:

> $\Delta E_p = mg\Delta h$ Δh is change in vertical height

EXAMPLE

Calculate the change in potential energy of a man of mass 80 kg who climbs a flight of stairs to increase his height above the ground by 20 m   ($g = 9.8$ m s⁻¹).

Using $\Delta E_p = mg\Delta h$
$$\Delta E_p = 80 \times 9.8 \times 20$$
$$\Delta E_p = 15\,680 \text{ J } (16 \text{ kJ 2 s.f.})$$

Power

Power (P) is equal to the rate at which work is done or the rate at which energy is being transferred.

$$P = \frac{\text{work done}}{\text{time taken}} = \frac{\Delta W}{\Delta t}$$

Power is measured in **watts** (W). If work is being done at the rate of 1 J s⁻¹, the power is 1 W.

EXAMPLE

If the man in the previous example takes 40 s to pull his boat 50 m, calculate his power.

$$P = \frac{\Delta W}{\Delta t}$$

$$P = \frac{4698.5}{40}$$

$$P = 117.5 \text{ W} \quad (120 \text{ W 2s.f.})$$

$$\text{Power} = \frac{\text{work done}}{\text{time taken}} = \frac{\text{force} \times \text{displacement}}{\text{time taken}} = \text{force} \times \text{velocity}$$

$$P = F \times v$$

> **EXAMINER'S TOP TIP**
> Work and energy are measured in joules (J). Power is measured in watts (W).

Force against velocity graph

The area under a force against velocity graph is equal to energy transferred each second.

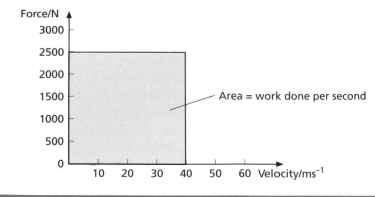

Area = work done per second

EXAMPLE

The diagram shows the force-time graph for a car travelling at a constant speed of 40 m s⁻¹. The propulsive force from the engine is 2500 N. Calculate the power of the engine.

$$P = F \times v$$
$$P = 2500 \times 40$$
$$P = 100\,000 \text{ W or } 100 \text{ kW}$$

Quick test

1 Calculate the work done when a force of 25 N moves an object 8.0 m in the direction of the applied force.

2 Calculate the work done when a force of 100 N, applied at an angle of 45° to the horizontal, is used to pull a trolley along a flat path for 20 m.

3 Calculate the kinetic energy of a bullet of mass 50.0 g travelling at a speed of 200 m s⁻¹.

4 Calculate the change in gravitational potential energy of a rock of mass 10.0 kg that falls through a vertical distance of 40 m ($g = 9.8$ m s⁻¹).

5 Calculate the power of an engine that does 100 kJ of work in 5 minutes.

6 A car engine develops a power of 30 kW when travelling at a constant speed of 50 m s⁻¹. Calculate the frictional forces experienced by the car.

1. 200 J   2. 1414 J   3. 1000 J   4. −3900 J   5. 333 W   6. 600 N

Energy sources and transfers

- Electricity is one of the most convenient forms of energy. It is easily converted into other forms of energy. Most of the electrical energy we use at home is generated at power stations. There are several different types of power station, but the most common in the UK use the fossil fuels, coal, oil or gas, as their source of energy.

- Fuel is burned to and <u>chemical energy</u> is released.

- <u>Heat energy released</u> heats water and turns it into steam.

- <u>Turbines</u> are turned by the steam.

- <u>Large generators</u> are turned by turbines.

- <u>Electrical energy</u> produced by generators.

- We receive electrical energy in our homes from the <u>National Grid</u>.

EXAMINER'S TOP TIP
Check your specification to see how much of the material from this spread you need. You may not need the information about different sources of energy.

Energy transfers

When energy is used it does not disappear. It is transferred into other forms.
A loudspeaker changes electrical energy into sound.
A light bulb changes electrical energy into heat and light energy.
As an object falls, some of its gravitational potential energy changes into kinetic energy.
The <u>law of conservation of energy</u> states that <u>energy cannot be created or destroyed</u>. The energy input into a device must be equal to the energy output.

Fossil fuels

<u>Fossil fuels</u> are a concentrated source of energy, but they are non-renewable fuels. Once they have been used up, they cannot be replaced. Their use also creates several environmental problems including the release of gases that contribute to the Greenhouse Effect, acid rain, and damage by mining or spillage during transport. Part of the solution to these problems is to use renewable sources of energy such as wind, waves, tidal, solar, geothermal, biomass and hydroelectric. Each of these sources has some advantages and disadvantages.

EXAMINER'S TOP TIP
Efficiency is usually expressed as a percentage but it can be expressed as a decimal, for example 0.75 efficient or 75% efficient.

Efficiency

Ideally, we would like all the electrical energy which enters a filament bulb to be changed into light energy. But this is not the case. A lot of the electrical energy is changed into unwanted heat.
To calculate the efficiency of a device/energy transfer we use this equation:

$$\text{Efficiency} = \frac{\text{useful energy output}}{\text{total energy input}} \times 100\%$$

In the light bulb example, 200 J of electrical energy enters the bulb, and 8.0 J of it are changed into light energy.

$$\text{Efficiency of bulb} = \frac{8}{200} \times 100\% = 4\%$$

A similar equation for efficiency can be written in terms of the total power input and the useful power output.

$$\text{Efficiency} = \frac{\text{useful power output}}{\text{power input}} \times 100\%$$

EXAMPLE
The motor of a crane has a power rating of 5.0 kW. Calculate its efficiency when it is doing useful work at the rate of 3.0 kW.

$$\text{Efficiency} = \frac{\text{useful power output}}{\text{power input}} \times 100\%$$

$$\text{Efficiency} = \frac{3.0}{5.0} \times 100\% = 60\%$$

Alternative sources of energy

Source	Advantages	Disadvantages
Wind power The kinetic energy of the wind is used to drive turbines and generators.	It is a renewable source of energy and therefore will not be exhausted. Has low-level technology and therefore can be used by developing countries. No atmospheric pollution. No fuel costs.	Visual and noise pollution. Limited to windy sites. No wind, no energy.
Hydroelectricity The kinetic energy of flowing water is used to drive turbines and generators.	Renewable source. Energy can be stored until required. No atmospheric pollution. No fuel costs.	High initial cost. High cost to environment, i.e. flooding, loss of habitat.
Wave power The rocking motion of waves is used to generate electricity.	Renewable source. No atmospheric pollution. Useful for isolated islands. No fuel costs.	High initial cost. Visual pollution. Poor energy capture. Large area of machines needed, even for small energy return.
Tidal power At high tide water is trapped behind a barrage or dam. When it is released at low tide the **gravitational potential energy of the water** changes into kinetic energy which then drives turbines and generates electricity.	Reliable, two tides per day. No atmospheric pollution. No fuel costs.	High initial cost. Possible damage to environment, for example flooding. Obstacle to water transport.
Geothermal In regions where the Earth's crust is thin, hot rocks beneath the ground can be used to heat water, turning it into steam. This steam is then used to drive turbines and generate electricity.	No fuel costs. No pollution and no environmental problems.	Very few suitable sites. High cost of drilling deep into the ground.
Biomass The energy in sunlight is stored as chemical energy in things that have grown, for example trees and plants.	Renewable source of energy. Low-level technology and therefore useful in developing countries. Does not add to the Greenhouse Effect as the carbon dioxide they release when burned was taken from the atmosphere as they grew.	Large areas of land needed to grow sufficient numbers of trees.
Solar energy The energy carried in the sunlight can be converted directly into electricity or absorbed by objects, increasing their thermal energy.	No pollution. No fuel costs.	Initially quite expensive. May not be so useful in regions where there is limited sunshine.
Nuclear The energy released during induced decay of nuclei is used to heat water, changing it into steam to drive turbines and generators.	Very small amounts of fuels used. No pollution if operated correctly.	Danger of major accidents. High decommissioning costs. High cost of treatment and storage of radioactive waste.

Quick test

1 Why is electrical energy often described as a 'convenient form of energy'?

2 Write down the names of two sources of energy which:

  a are intermittent **b** have a high initial cost **c** require only low technological skills.

3 Explain, using the law of conservation of energy, why water at the bottom of a waterfall is slightly warmer than the water at the top.

4 Calculate the efficiency of a radio which produces just 36 J of sound energy from 400 J of electrical energy.

5 Calculate the efficiency of an electric motor which has a power rating of 1.5 kW and does work at the rate of 1.2 kW.

1. It is easily converted into other forms. 2. a) wind and solar b) tidal and hydroelectric c) wind and biomass 3. Water at the top of a waterfall has gravitational potential energy. As it falls, it changes into kinetic energy. When this moving water hits the pool at the bottom of the fall, this kinetic energy does not disappear but is converted into heat energy (and some sound). 4. 9.0% 5. 80%

Circular motion

EXAMINER'S TOP TIP
Check your specification. You may not need to know about this topic for your AS exam.

Measuring angles

- If an object moves along the circular path from A to B it will travel a distance s.
- Its angular displacement is $\theta = s/r$ and is measured in radians.
- $360° \equiv 2\pi$ radians
- 1 radian = 57.3°.

EXAMPLE

Calculate the angular displacement for an object which:

a) travels 40 cm along the arc of a circle that has a radius of 20 cm

b) completes 3 laps of the circle.

a) $\theta = s/r = 40/20 = 2$ radians or 114.6°

b) Each time the object completes one lap of the circle, it will have an angular displacement of 2π radians. This object will therefore have an angular displacement of 6π radians.

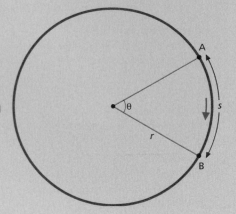

Angular velocity

- angular velocity = $\dfrac{\text{angular displacement}}{\text{time}}$

- $\omega = \dfrac{\theta}{t}$

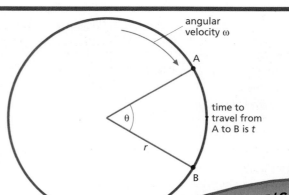

EXAMPLE

Calculate the angular velocity of a car that travels around $\frac{1}{4}$ of a circular race track in 10 s.

$$\omega = \frac{\theta}{t} = \frac{\pi/2}{10} = 0.16 \text{ rads s}^{-1} \text{ or } 9 \text{ degrees s}^{-1}$$

EXAMINER'S TOP TIP
Try to identify the similarities between linear motion and circular motion. For example, distance (s) and angle (θ), or linear velocity (v) and angular velocity (ω), etc.

Linear and angular velocity

Relationship between linear and angular velocity

If an object travels around the arc AB in t seconds, its angular displacement is $\dfrac{s}{r}$

If the object is moving with an angular velocity ω and it takes t seconds to travel from A to B, its angular displacement is ωt

therefore $\quad \dfrac{s}{r} = \omega t \quad$ or $\quad \dfrac{s}{t} = r\omega$

but $\quad \dfrac{s}{t} = v$ (linear velocity)

therefore the relationship between the linear and angular velocity is given by this equation:

$$v = r\omega$$

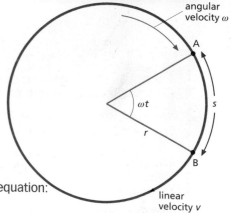

EXAMPLE

Calculate the angular velocity of a runner who travels at 10 m s^{-1} around a circular track of radius 250 m.

$v = r\omega \qquad$ therefore $\quad \omega = \dfrac{v}{r} = \dfrac{10}{250} = 0.04 \text{ rad s}^{-1} \text{ or } 2.3 \text{ degrees s}^{-1}$

Period and frequency

- The time it takes an object to complete one lap or one revolution is called the **period** (T).
- The number of laps or revolutions an object makes in one second is called the **frequency of the motion** (f).
- The frequency and period of an object rotating or moving in a circle are related by the equation $T = \dfrac{1}{f}$

EXAMPLE

Calculate the period of a disc rotating with a frequency of 600 r.p.m. (r.p.m. = revolutions per minute)

600 r.p.m. = 10 revolutions per second

Therefore $T = \dfrac{1}{10} = 0.1$ s.

Centripetal force and acceleration

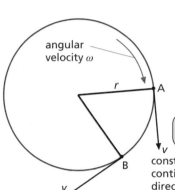

angular velocity ω

r

A

v

B

v

constant speed but continuously changing direction

An object travelling in a circle at a constant speed is accelerating because its direction is continually changing. The magnitude of this acceleration (a) is given by the equation

$$a = r\omega^2 \text{ or } a = \frac{v^2}{r}$$

The direction of the acceleration is always towards the centre of the circle.

In accordance with Newton's second law, if there is an acceleration there must be a force causing it. This force is called the **centripetal force** and acts towards the centre of the circle. If the mass of the object is m then, from F = ma, the size of the centripetal force is given by this equation:

$$F = mr\omega^2 \text{ or } F = \frac{mv^2}{r}$$

Examples of the different kinds of forces that can cause an object to travel in a circle include:

- gravitational forces which cause planets, moons and satellites to travel in orbits
- electrostatic forces which cause electrons to orbit nuclei
- tensions in rods, wires and strings; for example throwing the hammer
- friction and reaction forces.

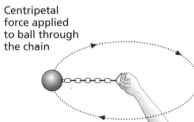

Centripetal force applied to ball through the chain

NOTE: The force is always applied towards the centre of the circle. The ball therefore accelerates towards the centre of the circle.

If a centripetal force disappears, for example the chain snaps, the object will instantly stop moving in a circle and will instead travel in a straight line and at a constant speed. Its path will be the tangent to the circular path at the point where the force disappeared.

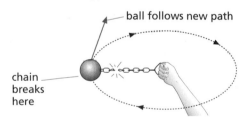

ball follows new path

chain breaks here

Quick test

1 Calculate the angular displacement of a circling aircraft which travels 3 km along the arc of a circle which has a radius of 10 km.

2 Calculate the angular velocity of a car travelling around a circular race track at 200 m s⁻¹. The radius of the track is 500 m.

3 Calculate the linear velocity of a runner who has an angular velocity 0.14 rad s⁻¹ whilst travelling around a track of radius 63 m.

4 Calculate the frequency of a spinning top that has a period of 0.002 s.

5 A ball of mass 400 g is made to travel in a horizontal circle of radius 1.5 m. If the linear speed of the ball is 10 m s⁻¹, calculate the following:

   a the acceleration of the ball towards the centre of the circle

   b the centripetal force applied to the ball by the string.

Exam-style questions
Use the questions to test your progress. Check your answers on pages 92–95.

1 A car of mass 1000 kg and travelling at 25 m s^{-1} collides with a stationary car of mass 1500 kg.
  a Assuming that both cars remain in contact after the collision, calculate their speed. [4]

....

....

....

  b Determine through calculation whether this is an elastic or inelastic collision. [4]

....

....

....

2 a A man of mass 80 kg steps off a stationary boat of mass 120 kg. If the velocity of the man is
  3 m s^{-1} eastwards, determine the velocity of the boat. [2]

....

....

  b Explain briefly the principle of rocket propulsion. [4]

....

....

....

3 A ball of mass 200 g, travelling at a speed of 10 m s^{-1}, is struck by a bat. The ball subsequently
  travels back along its original path and at the same speed.
  a Calculate the change in velocity of the ball. [1]

....

  b Calculate the change in momentum of the ball. [1]

....

  c If the ball was in contact with the bat for 0.05 s, calculate the average force applied to the
  ball during this time. [2]

....

....

4 The diagram shows an object of mass 4.0
  kg supported in the hand of an extended
  forearm.
  The forearm is 0.4 m long and pivots about
  point X. Its centre of gravity is 0.20 m from
  X and its effective mass is 1.5 kg. The
  forearm and the object are supported by
  an upward force F which is provided by
  the biceps muscle and acts 0.050 m from
  X. ($g = 9.8$ m s^{-2})
  Calculate the following magnitude of the
  force F.

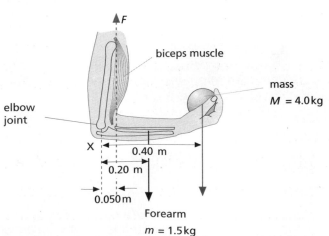

biceps muscle

mass
$M = 4.0\,kg$

elbow
joint

X

0.40 m

0.20 m

0.050 m

Forearm
$m = 1.5\,kg$

.... [4]

....

....

....

5 A stone of mass 400 g is dropped vertically from a cliff. It strikes the ground at a speed of 40 m s^{-1}. Calculate the following: (g = 9.8 m s^{-2})

  a The kinetic energy of the stone just before it hits the ground. [2]

....

....

  b The gravitational potential energy of the stone before it is released. [1]

....

  c The height of the cliff. (g = 9.8 m s^{-2}) [2]

....

....

6 An athlete of mass 70 kg runs up a 500 m hill in 5 minutes.
Calculate the following. (g = 9.8 m s^{-2})

  a The increase in the gravitational potential energy of the athlete. [2]

$mgh = 70 \times 500 \times 9.8$

$= 343,000\ J \qquad - 343\ kJ$

  b The average power of the athlete during his run. [2]

$Power = \dfrac{work\ done}{time} = \dfrac{343\,000}{}$

$= 1143\ -\ W$

  c Explain why the energy used by the runner during the climb will be much greater than his increase in gravitational potential energy. [1]

....

7 a Calculate the power of a pump which lifts 240 kg of water per minute through a height of 4.0 m [3]

$\dfrac{240 \times 9.8 \times 4}{60} = 9408 \quad = 156.8\ W$

....

....

  b Calculate the power of a car engine which, when it is travelling at its maximum speed of 60 m s^{-1}, is producing a propulsive force of 5.0 kN. [2]

$P = F \times V = 5000 \times 60$

$= 300,000\ W \quad or\ 300\ kW$

8 a Calculate the angular velocity of the second hand on a watch. [2]

....

....

  b Calculate the linear speed of the point of a second hand that is 1.5 cm long. [1]

....

Total   /40

Basic wave properties

- Waves transfer energy without transferring matter.
- All waves consist of vibrations or oscillations.

EXAMINER'S TOP TIP
Make sure you understand all the basic properties and the names of the principal wave features. You will need to use them often in this topic.

Transverse and longitudinal waves

There are two main types of waves. These are **transverse waves** and **longitudinal waves**.

A transverse wave has vibrations of the medium that are at **right angles** to the direction in which the wave is moving.

A longitudinal wave has vibrations that are **along** the direction in which the wave is moving.
Sound waves are longitudinal waves.

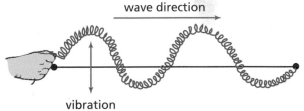

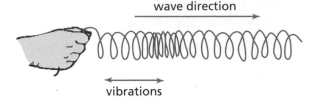

Examples of transverse waves include electromagnetic waves and surface water waves.

Waves like the ones above can be seen to move along a slinky. They are called **progressive waves**.
Stationary waves have a profile or shape that does not move through the medium (see page 40).

The important bits

Wavelength (λ)
the length of one complete cycle, for example the distance between the peak of one wave and the peak of the next.

Amplitude (a)
the maximum displacement from the mean position.

Frequency (f)
the number of complete oscillations or vibrations of a particle each second. It is measured in hertz (Hz), where 1 Hz is one oscillation per second.

Period (T)
the time taken for one complete oscillation. The frequency and period of a wave are related by the equation
$$T = \frac{1}{f}$$

Speed (v)
the distance travelled by a wave in 1 s. The frequency, wavelength and speed of a wave are related by the equation
$$v = f\lambda$$

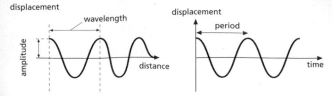

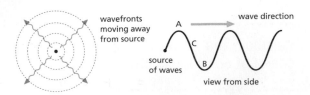

If a single small pebble is dropped onto the surface of a calm pond, it will create disturbances that we see as circular ripples or wave fronts. The distance between one wave front and the next is equal to one wavelength. Seen from the side, the water is vibrating up and down as the wave energy is moving away from the source. All points on the same wave front are at the same part of their oscillatory motion. These points are said to be vibrating in **phase**. Particles that are not on the same wave front may be at different parts of their oscillatory motion.

Points A, B and C are not vibrating in phase. They have a **phase difference**. We can express this difference as an angle.

The phase difference between points A and B is 180°.

The phase difference between A and C is 90°.

Note A phase difference equal to a multiple of 360° indicates that the vibrations of the particles are in phase.

Power and wave intensity

As a wave front moves away from its point of origin, the energy it carries becomes more dilute, i.e. it is spread over a larger area. The intensity at a point is the energy arriving per second per square metre.

In the case of a point source for electromagnetic waves or sound waves, the wave fronts will be spherical and the intensity I at a distance r from the source can be calculated using the equation

$$I = \frac{P}{4\pi r^2}$$

where P is the energy emitted by the source each second (power). Note that this is an inverse square law, i.e.

$$I \propto \frac{1}{r^2}$$

EXAMPLE

The power of a point source of sound waves is 10 W. Calculate the intensity at a distance of 2.0 m from the source.

$$I = \frac{P}{4\pi r^2}$$

$$I = \frac{10}{4\pi \times 2.0^2}$$

$$I = 0.2 \ \textbf{Wm}^{-2}$$

- Some of the energy being carried by a wave may be lost to the medium through which it is travelling. As a result the temperature of the medium may rise.
- The intensity I depends upon the amplitude A of vibration, at that point. $I \propto A^2$.

Quick test

1 Explain the difference between a transverse wave and a longitudinal wave.

2 What are the amplitude and wavelength of this wave?

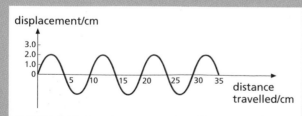

3 What are the period and the frequency of this wave?

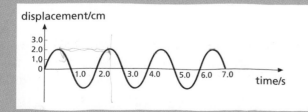

4 Calculate the frequency of red light that has a wavelength of 690 nm.
($c = 3.0 \times 10^8$ m s^{-1})

5 Which two points on this wave are vibrating

a in phase

b 90° out of phase

c 180° out of phase?

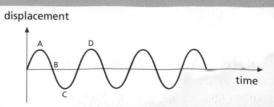

6 The total energy emitted by the Sun each second is approximately 4.0×10^{26} J. Calculate the intensity of the light reaching the Earth, given that the distance from the Sun to the Earth is 1.5×10^{11} m.

1. The vibrations of a transverse wave are at 90° to the direction in which the wave is moving. For a longitudinal wave the vibrations are along the direction in which the wave is moving. 2. 2.0 cm and 10 cm 3. 2.0 s and 0.5 Hz 4. 4.3×10^{14} Hz 5. a) A and D b) A and B or B and C c) A and C or C and D 6. $I = 4 \times 10^{26}/4 \times 3.14 \times (1.5 \times 10^{11})^2$, approximately 1.4 kWm^{-2}

31

Reflection and refraction

Reflection

Mirrors form images because they reflect light. When reflection occurs the **angle of incidence is equal to the angle of reflection**.

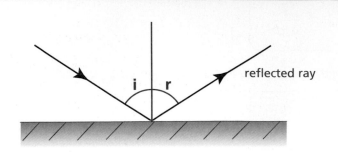

reflected ray

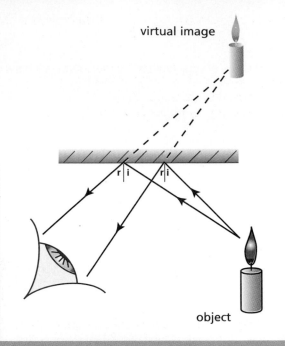

virtual image

object

A plane mirror creating an image.

EXAMINER'S TOP TIP
Check the topics covered on this spread with the specification for the exam you will be taking. Your specification may not require you to know all this material.

Refraction

If a wave travels across the boundary between two media its speed may change and, as a result, its direction may also change. This change in direction is called **refraction**.

If the speed of a wave decreases as it crosses a boundary it will be refracted towards the normal. If the speed of the wave increases it will be refracted away from the normal.

ray bends towards normal as it enters the glass

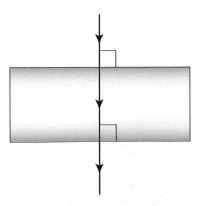

normal

air   glass

normal

ray bends away from the normal as it leaves the block (parallel to original ray).

A wave striking the boundary at 90° is not refracted.

The amount by which a wave changes its direction as it crosses a boundary depends on:
- the wavelength of the wave – the longer the wavelength, the smaller the deviation
- the change in the speed of the wave as it crosses the boundary – the greater the change, the greater the deviation. This change is described by the **refractive index** of the material.

Refractive index

The **absolute refractive index** of a material (n) is the ratio of the speed of a wave in a vacuum (c) compared with the speed of that wave in the medium (c_1)

$$n = \frac{c}{c_1}$$

If neither of the media is a vacuum the equation becomes:

$$_1n_2 = \frac{c_1}{c_2}$$

$_1n_2$ comparative refractive index
c_1 speed of wave in medium 1
c_2 speed of wave in medium 2

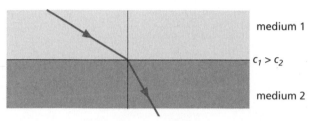

medium 1

$c_1 > c_2$

medium 2

Note The refractive index of a vacuum and the refractive index of air are usually taken as 1.

EXAMPLE

A ray of light travels through a piece of glass at a speed of 2.0×10^8 m s^{-1}.
If the speed of light in air is 3.0×10^8 m s^{-1}, calculate the absolute refractive index of the glass.
Using
$$n = \frac{c}{c_1} = \frac{3.0 \times 10^8}{2.0 \times 10^8} = 1.5$$

The refractive index is also related to the angle of incidence (θ_1) and the angle of refraction (θ_2) of a wave as it crosses the boundary between two media by this equation:

$$_1n_2 = \frac{\sin \theta_1}{\sin \theta_2}$$ This is known as **Snell's law**.

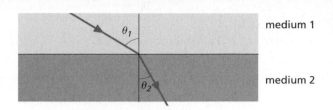

medium 1

θ_1

θ_2

medium 2

The refractive index is also related to the critical angle of the medium by this equation:

$$n = \frac{1}{\sin c}$$ c is the critical angle.

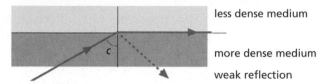

less dense medium

c

more dense medium

weak reflection

EXAMPLE

Calculate the angle of refraction (θ_2) for the ray of light in the diagram below, and the critical angle for the air/plastic boundary.

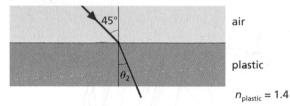

45°

air

θ_2

plastic

$n_{plastic} = 1.4$

Using
$$_1n_2 = \frac{\sin \theta_1}{\sin \theta_2}$$

$$\sin \theta_2 = \frac{\sin \theta_1}{_1n_2} = \frac{\sin 45}{1.4} = 0.505$$

$$\theta_2 = 30.3°$$

Using $$n = \frac{1}{\sin C}$$

$$\sin C = \frac{1}{n}$$

$$\sin C = \frac{1}{1.4} = 0.714$$

$$C = 45.6°$$

Quick test

1 Name two factors which affect how much a wave changes direction as it crosses the boundary between two media.

2 The speed of light in glass A is 2.1×10^8 m s^{-1}. The speed of light in glass B is 1.9×10^8 m s^{-1}. Calculate the refractive index for a ray travelling from glass A to glass B.

3 Calculate the absolute refractive index and the critical angle for the plastic in the diagram.

4 Calculate the angle of refraction for a ray that strikes the surface of a block of glass at an incidence angle of 45°. The absolute refractive index of the glass is 1.3.

5 Assuming that the speed of light in a vacuum is 3.0×10^8 m s^{-1}, calculate the speed of the ray in question 4 as it passes through the glass block.

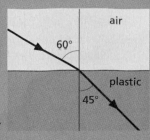

air

60°

plastic

45°

1. The change in the speed of the wave as it crosses the boundary and the wavelength of the wave. 2. 1.1 3. 1.2, 55°
4. 33° 5. 2.5×10^8 m s^{-1}

Total internal reflection and polarisation

- The inner surface of a glass block can sometimes behave like a mirror. This is called <u>total internal reflection</u>.
- Total internal reflection happens if the ray strikes the inside surface at an angle greater than the <u>critical angle</u>.
- If the angle of incidence is small, the ray is refracted as usual.
- If the angle of incidence is large then total internal reflection takes place.
- Total internal reflection only occurs when light travels from a more optically dense material into a less optically dense material (for example, glass to air).

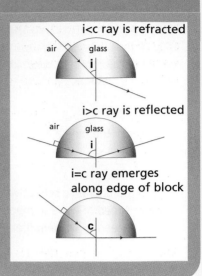

i<c ray is refracted

i>c ray is reflected

i=c ray emerges along edge of block

Total internal reflection in prisms

Turning a ray through 90°

This is used in prismatic periscopes.

Turning a ray through 180°

The critical angle for glass is approximately 42°.

This is used in bicycle reflectors and binoculars.

Optical fibres and light pipes

- An optical fibre has a **high-density glass for its core** and a **less dense (lower refractive index) glass as an outer coating**.
- The fibre is so narrow that light entering at one end will always strike the boundary between the two glasses at an angle greater than the **critical angle**. It will therefore undergo a **series of total internal reflections** before emerging at the far end of the fibre.
- Fibres can be bundled together to make **light pipes**.
- Because these pipes are flexible we can use them to see and bend light round corners.
- Optical fibres are now used to carry signals, for example cable TV.

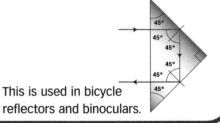

more dense glass   less dense glass

Advantages of optical fibres over copper wire and cables in telecommunications

Optical fibres are gradually being used to replace copper wires and cables in telecommunications. They have several advantages over copper wires.

- Optical fibres are capable of carrying far more information. Just one fibre could carry as many as 30 000 phone calls.
- Optical fibres are free from noise.
- There is little loss in energy so signals can travel far greater distances along optical cables without needing regeneration (100 km in optical fibres compared with 4 km in copper cables).
- Optical fibres are much lighter and smaller.
- It is much more difficult to tap into or bug a conversation that is being transmitted through optical cables.

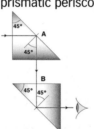

lens creates image of inside of body   light returns from inside body

light travels down fibres into body

light from light source

The endoscope

- Used by surgeons to see inside the body.
- Light travels down one set of fibres and is reflected back through another set.

Polarisation

Many transverse waves, including electromagnetic waves, have vibrations in all planes that are at right angles to the direction of travel. Such waves are said to be **unpolarised**. If some of these vibrations are removed so that there is vibration in one plane at right angles to the direction of travel, the waves are said to have been **polarised**.
Light waves can be polarised using a sheet of polarising filter.

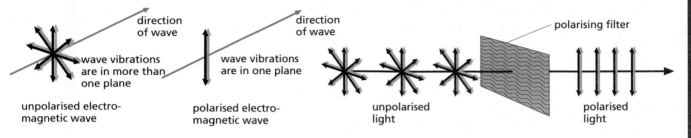

Radio waves are polarised at transmission. In certain areas where reception is poor, the signal may be boosted by a local relay station before being re-transmitted. To prevent interference between the original signal and the boosted signal, they are transmitted such that one is polarised horizontally and one is polarised vertically.

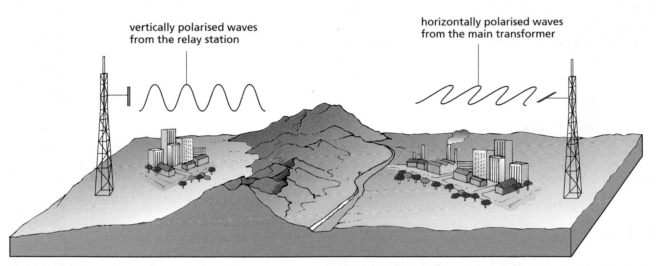

Note Only transverse waves can be polarised. The vibrations of longitudinal waves are along the direction in which they are travelling. Longitudinal waves therefore cannot be polarised.

Quick test

1 Explain the conditions necessary for a ray of light to undergo total internal reflection.

2 Explain why light travelling in an optical fibre never escapes through the walls.

3 Give four reasons why optical fibres are used to replace copper wires in the telecommunications industry.

4 What is an endoscope? How does it make use of optical fibres?

5 Explain the difference between a polarised and an unpolarised transverse wave.

6 Why can longitudinal waves not be polarised?

1. The ray must be travelling from a more dense medium to a less dense medium and strike the boundary at an angle greater than the critical angle. 2. Light travelling within the core always strikes the boundary between the two glasses at an angle greater than the critical angle and therefore is totally internally reflected and not refracted. 3. Optical fibres are capable of carrying far more information. Optical fibres are free from noise. There is little loss in energy so signals can travel far greater distances along optical cables before needing regeneration. Optical fibres are much lighter and smaller. It is much more difficult to tap into or bug conversations that are being transmitted along optical fibres. 4. An endoscope is an optical instrument used by doctors to see inside the body. Optical fibres carry light into the body to illuminate it and they carry the reflected light back to an eyepiece so that the body part can be seen. 5. The vibrations of an unpolarised wave are in all planes at right angles to the direction of travel. The vibrations of a polarised wave are in just one plane at right angles to the direction of travel. 6. Their vibrations are along the direction in which the waves are moving.

Diffraction

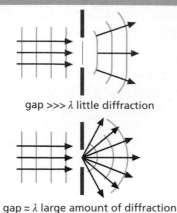

gap >>> λ little diffraction

gap ≈ λ large amount of diffraction

- **Diffraction** is the spreading out of waves as they pass through a gap or the edge of an obstacle. Diffraction is most noticeable if the size of the gap is approximately equal to the wavelength of the waves being observed.

- The wavelength of a sound wave may be similar to the width of a doorway. As a result, waves can be diffracted through large angles as they pass through.

Diffraction of a sound wave

EXAMINER'S TOP TIP
Remember diffraction may not be observable if the size of the gap and the wavelength of the waves are very different.

Diffraction of radio waves

Without diffraction, radio waves could only be received by places that are in line of sight of the transmitter. The signal would therefore have a very limited range. Medium and long wavelength radio waves are diffracted by the Earth's surface. As a result, they can be received in places that are not in line of sight of the transmitters. Diffraction has increased the range of these signals.

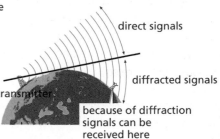

receiver – just in line of sight and so will receive signals without diffraction

transmitter

Earth

no signals will be received here

direct signals

diffracted signals

transmitter

because of diffraction signals can be received here

Diffraction pattern

The photograph below shows the diffraction pattern created by a single slit illuminated by monochromatic light.

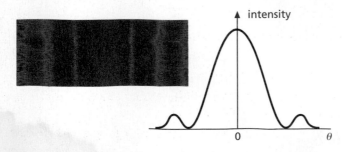

intensity

0   θ

The photograph clearly shows a central bright area with several dark and light fringes on either side. The graph shows how the intensity of the light varies as we move outwards from the central maximum.

This pattern has been created because rays of light from different parts of the slit have travelled different distances. Where the path difference between these waves is zero or whole number of wavelengths, constructive interference

will take place. Where the path difference is half a wavelength or 1½ wavelengths etc., destructive interference will take place and a minimum will be seen at these places.

path difference between these two waves = $\frac{a}{2}\sin\theta$

A

a

B

θ

θ

If the size of the gap is a and the wavelength of the light is λ, the first minimum will be seen at an angle of θ where

$$\sin\theta = \lambda / a$$

From the above, we can see that decreasing the size of the gap will increase the width of the central maximum but, because less light energy is passing through the gap, the intensity of the pattern as a whole will be less.

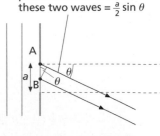

intensity

pattern produced by wide slit

pattern produced by narrower slit

0   θ

Diffraction pattern continued

EXAMPLE

Calculate the angle between the centre of the diffraction pattern and the first minimum when light of wavelength 450 nm passes through a single slit 0.15 mm wide.

Using $\sin \theta = \lambda / a$

$\sin \theta = 450 \times 10^{-9} / 0.15 \times 10^{-3}$

$\sin \theta = 0.003$

$\theta = 0.17°$

The diffraction grating

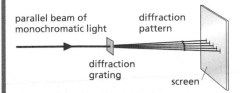

parallel beam of monochromatic light

diffraction pattern

diffraction grating

screen

A **diffraction grating** is a piece of glass on which lines or grooves have been drawn in order to produce multiple slits. When monochromatic light is shone onto the grating, a large number of coherent light sources are created. Waves from these sources diffract and overlap, creating an interference pattern. Although there are some similarities between this experiment and Young's double slit experiment (see page 39), there are several important differences.

- Because the slit separations are very small, the maxima produced by the grating are more widely spaced.
- Because there are more slits the maxima are much brighter.
- Because there are a large number

of slits, constructive interference only takes place within a very narrow angle, hence the maxima are much sharper.

Consider two rays emerging from the same position in adjacent slits.

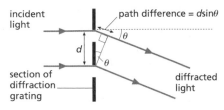

incident light

path difference = $d\sin\theta$

d

section of diffraction grating

diffracted light

For **constructive interference** to take place, the path difference between these waves must be a whole number of wavelengths. From the geometry of the arrangement we can see that the path difference between the two waves is $d \sin \theta$.

For constructive interference to occur we can therefore write the general equation

$$d\sin \theta = n\lambda$$

where
- d the slit separation
- θ the angle to the straight on direction in which constructive interference is taking place
- n the number or order of the maximum being observed.

Note The maximum directly in front of the slit is known as the zero order maximum ($n = 0$) as the path difference between the waves is zero. Using visible light it is possible to obtain three or four maxima.

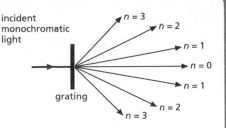

incident monochromatic light

grating

$n = 3$
$n = 2$
$n = 1$
$n = 0$
$n = 1$
$n = 2$
$n = 3$

EXAMPLE

Calculate the angular position of the third order maximum created when monochromatic light of wavelength 600 nm passes through a diffraction grating with a slit separation of 2.0×10^{-6} m.

Using

$d\sin \theta = n\lambda$

$\sin \theta = n\lambda/d$

$\sin \theta = \dfrac{3 \times 600 \times 10^{-9}}{2.0 \times 10^{-6}}$

$\sin \theta = 0.900$

$\theta = 64°$

Note Diffraction gratings are often described by a number (N) which indicates the number of lines drawn on each metre or millimetre of the grating. To obtain the value for d, this number must be expressed in metres and then its reciprocal found.

i.e. $d = 1/N$

EXAMPLE

Calculate the distance between adjacent slits for a diffraction grating that has 10^2 lines per mm.

If the grating has 10^2 lines per mm, it will have 10^5 lines per m.

$$d = \frac{1}{N}$$

Therefore $d = 1.0 \times 10^{-5}$ m

Quick test

1 What type of radio wave is diffracted most by the Earth's surface? How does this affect the possible range of these signals?

2 Explain how waves from a single slit can produce an interference pattern.

3 Calculate the width of a single slit which, when illuminated with light of wavelength 700 nm, creates a minimum at an angle of 8.0° to the straight on direction.

4 Describe two changes that take place to the diffraction pattern created by a single slit if the width of the slit is halved.

5 Describe three differences between the interference pattern created in Young's two slit experiment and that created by a diffraction grating.

6 Calculate the distance between each slit for a diffraction grating that has 50 lines per mm.

7 Calculate the angular positions of the second and third order maxima when light of wavelength 750nm is incident normally on a diffraction grating with a slit separation of 2.5×10^{-6} m.

1. long wave (and medium wave) The ranges of these waves are increased. 2. There is a path difference between overlapping waves that have come from different parts of the slit. 3. 0.0050 mm 4. The intensity of the pattern is less and the angle where the first minimum occurs is larger. 5. The maxima created by a diffraction grating are more widely spaced, brighter and sharper. 6. 2.0×10^{-5} m 7. 37°, 64°

37

Diffraction

Interference

The principle of superposition

When two or more sets of waves pass through the same point, the resultant displacement at a given time is the sum of the displacements of each wave.

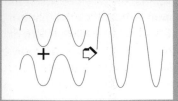

- When waves arrive at a point in phase, a large amplitude of vibration is produced. This is called <u>constructive interference</u>.
- When two waves arrive at a point in anti-phase, the resultant amplitude of vibration is a minimum. This is called <u>destructive interference</u>.

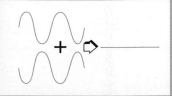

- In the two examples shown here, the two waves have the same frequency and amplitude. Therefore the resultant wave produced in the first example will have an amplitude twice the amplitude of one of the waves and the resultant in the second example will have an amplitude of zero, i.e. there is no resultant vibration.

EXAMINER'S TOP TIP
The creation of an interference pattern is direct proof that light travels as waves.

Interference patterns

The photo and diagram show the pattern created in a ripple tank when waves from two point sources overlap. It is called an **interference pattern**.

The resultant produced at any one point in the ripple tank depends upon the phase difference between the two waves at that point. If the waves arrive in phase the waves will interfere constructively and a maximum will be created. Assuming that the waves were emitted in phase, they will interfere constructively at all those points where the path difference between the waves is zero, 1λ, 2λ, etc.

At those points where the path difference between the waves is $1/2\lambda$, $3/2\lambda$, etc. the waves will arrive in anti-phase and minima will be created at these points.

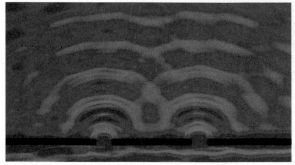

path difference = 0
path difference = $\frac{\lambda}{2}$
path difference = λ

——— constructive interference

- - - - destructive interference

EXAMINER'S TOP TIP
Examiners are keen that you understand what coherent waves are and why they are necessary to create sustained interference patterns. So read this section carefully.

Observing interference patterns

For an interference pattern to be observed, the sets of waves must:
- have approximately the same amplitude
- vibrate in the same plane
- be **coherent**

Waves are coherent or have coherent sources if:
- they have the same frequency and
- they maintain a constant phase relationship.

These waves are out of phase but they are coherent.

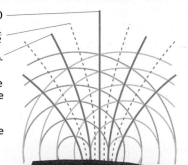

coherent wave sources

These waves are not coherent. The phase relationship between them is not constant.

sudden change in phase

incoherent wave sources

sudden change in phase

Interference patterns continued

Light waves from two separate sources are not coherent. Light is emitted in short bursts with no fixed phase relationship between one burst and the next. The abrupt changes in phase mean that it is not possible to create an observable interference pattern between light waves from different sources.

To create an observable interference pattern with light, it is necessary to create two light sources from a single light source. We can demonstrate this using Young's double slit experiment.

Young's double slit experiment

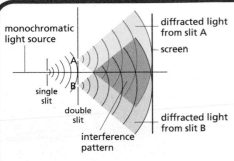

- Monochromatic light falls on a single slit creating a small light source.
- Light from this slit is diffracted and falls on two small slits creating two coherent sources.

- Light is diffracted from each of the slits.
- Where the light overlaps on the screen, an interference pattern can be seen.

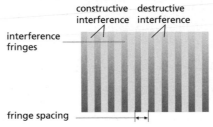

The relationship between the fringe spacing (W), the wavelength of the light (λ), the distance between the double slits and the screen (D) and the distance between the double slits (d) is given by the equation.

$$W = \frac{\lambda D}{d}$$

EXAMPLE

Two coherent sources 5.0 m apart emit sound waves of wavelength 1.4 m. Calculate the distance between successive maxima at a distance 7.0 m from the midpoint between the speakers.

$$W = \frac{\lambda D}{d} = \frac{1.4 \times 7.0}{5.0} = 1.96\,\text{m}$$

Interference by reflection

The direct waves and the reflected waves in the diagrams have travelled different distances to the receiver. If the path difference between them is a whole number of wavelengths, the waves will arrive in phase, will interfere constructively and a strong signal will be detected. If the path difference is an odd number of half wavelengths, the waves will arrive 180° out of phase, interfere destructively and a very weak signal will be detected. The height of the ionosphere above the Earth changes. As it does so, the path difference between the direct and refracted signals will therefore also change.

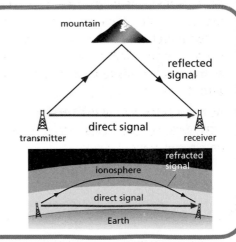

Quick test

1 What is constructive interference?
2 What is destructive interference?
3 What are coherent waves?
4 Why is it not possible to create an interference pattern using two separate light sources?
5 What proof do you have that light travels as waves?
6 Write down the Young's double slit equation.

1. When two or more waves overlap to create a resultant with a larger amplitude than the individual waves. 2. When two or more waves overlap to create a resultant with a smaller amplitude than the individual waves. 3. Coherent waves have the same frequency and a constant phase relationship. 4. Light waves from two separate sources are not coherent and therefore will not produce an observable interference pattern. 5. Light waves can be diffracted. They can also interfere. Both are properties of waves. 6. $W = \lambda D/d$

Standing or stationary waves

● The profile of a surface water wave moves with time. It is an example of a <u>progressive wave.</u>

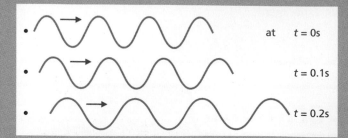

at $t = 0$ s

$t = 0.1$ s

$t = 0.2$ s

When these waves interfere there are several possible wave shapes that can be produced. The pattern corresponding to the lowest frequency is know as the <u>fundamental mode of vibration.</u>

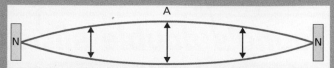

The <u>second mode of vibration</u> is called the <u>first overtone</u>.

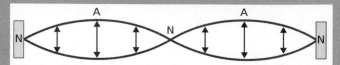

The <u>third mode of vibration</u> is called the <u>second overtone</u>.

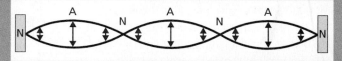

● Standing or stationary waves have a profile that does not move with time.

Stationary waves are created when two waves of the same wavelength and frequency travelling in opposite directions overlap. This is easily demonstrated using a vibrating string.

If a progressive wave (series of pulses) is sent along the length of string shown below, it is reflected when it reaches the fixed end.

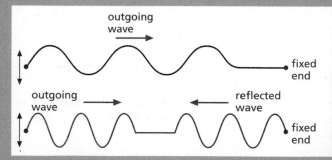

outgoing wave

fixed end

outgoing wave

reflected wave

fixed end

The outgoing wave and the reflected wave overlap and interfere, creating a wave pattern that has crests and troughs that do not move. This is a <u>standing wave</u>.

● There are points along the strings where there is no displacement at any time. These points are called <u>nodes</u>.
● There are points along the string where the amplitude of vibration is at a maximum. These points are called <u>antinodes</u>.
● The positions of the nodes and antinodes of a standing wave do not change with time.
● The distance between successive nodes or successive antinodes is $\lambda/2$.
● All particles between two adjacent nodes vibrate in phase, but the amplitude of vibration varies with position.

Frequency of vibration – strings

The frequency, f, of the fundamental mode of vibration of a string depends upon:
● the length of the string (l)
● the tension in the string (T)
● the mass per unit length of the string (μ)
The relationship is described by the equation $f = \frac{1}{2l} \sqrt{\frac{T}{\mu}}$

The frequency of vibration of any of the other modes is described by the equation

$$f = \frac{n}{2l} \sqrt{\frac{T}{\mu}}$$

where n is the number of loops or half wavelengths in the standing wave.

EXAMPLE
Calculate the mass per unit length of a string 0.5 m long, experiencing a tension of 50 N and vibrating in its third mode of vibration with a frequency of 100 Hz.

Using $f = \frac{n}{2l} \sqrt{\frac{T}{\mu}}$

$100 = \frac{3}{2 \times 0.5} \sqrt{\frac{50}{\mu}}$

$\mu = 0.045\,\text{kg m}^{-1}$

Standing waves in air

It is possible to create standing waves in air. A sound wave travelling from the open end of a pipe will be reflected when it reaches the closed end. The outgoing waves and the reflected waves then interfere to create standing waves in the air. As with the vibrations in a string, the air can vibrate in several different modes.

antinode

node

Fundamental mode (lowest frequency)

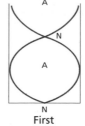

A

N

A

N

First overtone

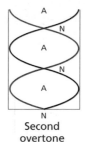

A

N

A

N

A

N

Second overtone

Determining the speed of sound using standing waves

- The time-base of the CRO is turned off. Sounds entering the microphone create a vertical trace, the height of which indicates the amplitude of vibration of a wave at that point.

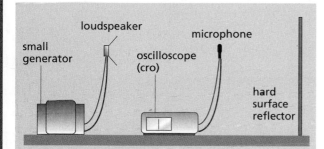

small generator

loudspeaker

oscilloscope (cro)

microphone

hard surface reflector

- The loudspeaker emits sound waves.
- The reflector is moved slowly back and forth until the trace seen on the CRO is a minimum.
- Reflected waves are combining with the waves from the loudspeaker to produce a standing wave. The microphone is at one of the nodes of this wave.
- The microphone is moved directly towards, or away from, the reflector until the position of the next minimum is found. The distance, d, between these minima is $\lambda/2$.
- If the frequency of the sound (f) emitted by the loud-speaker is known, the speed of sound in free air can be calculated using the equation $v = f \times \lambda$ or $v = f \times 2d$.

Quick test

1 Explain the difference between a progressive wave and a standing wave.

2 What three conditions are necessary for two waves to overlap and produce a standing wave?

3 Explain the difference between a node and an antinode.

4 What is the phase relationship between all points between two successive nodes of a standing wave?

5 Calculate the frequency of vibration for a piece of string 1.0 m long, having a mass per unit length of 0.02 kg m⁻¹ and vibrating in its fundamental mode when a tension of 200 N is applied to it.

6 What would be the frequency of vibration of the string in question 5 if it were vibrating

a in its second mode of vibration

b in its fourth mode of vibration?

7 A loudspeaker emits sound waves with a frequency of 170 Hz. Interference with reflected waves creates a standing wave. If the speed of sound in air is 340 m s⁻¹, calculate the distance between successive nodes in this standing wave.

1. Progressive wave → transfer energy. Standing wave → no energy transfer (localised energy)   2. The waves must have the same frequency, the same amplitude and be travelling in opposite directions.   3. A node is a point on a standing waves where there is no vibration. An antinode is a point where there is maximum amplitude of vibration.   4. All points between successive nodes on a standing wave vibrate in phase.   5. 50Hz   6. a) 100Hz b) 200Hz   7. 1.0 m

41

Exam-style questions
Use the questions to test your progress. Check your answers on pages 92–95.

1 a Explain the difference between transverse waves and longitudinal waves. Give one example of each. [4]

....
....
....
....

b Calculate the frequency of a wave that has a period of 2.0×10^{-2} s. [2]

....
....

c Calculate the wavelength of a sound wave with a frequency of 200 Hz.
(Speed of sound = 340 m s^{-1}) [2]

....
....

2 a i Sketch a displacement – time graph for a wave which has a displacement of zero at time t = 0. The graph should show two complete waves. [2]
  ii Label the amplitude and wavelength of the wave on your graph. [2]
  iii On the same axes, draw a second wave of identical wavelength and frequency, but with a phase difference of 90°. [2]

b A 100 W light bulb illuminates a room. Assuming that the bulb is 10% efficient, and that light energy is radiated equally in all directions, calculate the intensity of the light on a wall 2.0 m from the bulb. [4]

....
....
....
....

3 The diagram shows a ray of light travelling from water into glass.

a Explain why the ray of light changes direction as it crosses the boundary. [2]

....
....

b The absolute refractive index of water is 1.3.
  i Calculate the speed of light in water. [2]

....
....

  ii Calculate the comparative refractive index for light travelling from water into glass. [2]

....
....

  iii Calculate the speed of light in glass. [2]

....
....

4 The diagram below shows a ray of light travelling in an optical fibre.

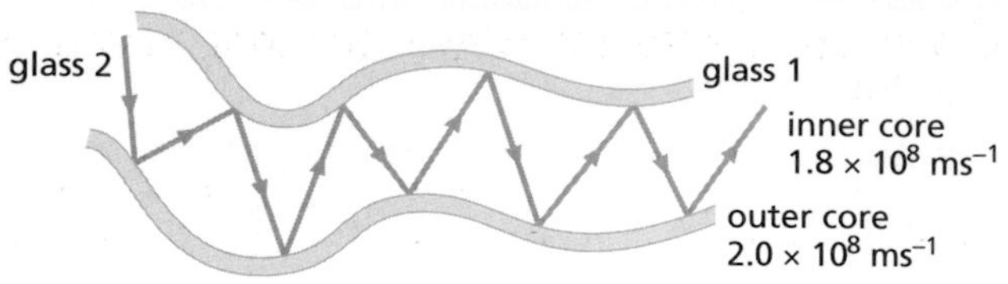

glass 2

glass 1
inner core
1.8×10^8 ms^{-1}
outer core
2.0×10^8 ms^{-1}

a Calculate the comparative refractive index $_1n_2$ for the two glasses. [2]

....

....

b Calculate the critical angle at this boundary. [2]

....

....

c Give three reasons why optical fibres are replacing copper wires in the telecommunications industry. [3]

....

....

....

5 a Explain the phrase 'coherent sources of waves'. [2]

....

....

b Explain why a man walking in front of a pair of speakers connected to the same signal generator hears a note whose loudness changes:
  i when he walks in a direction which is parallel to a line joining the two speakers [4]

....

....

....

....

  ii when he walks away in a direction which is at 90° to the line joining the two speakers. [2]

....

....

6 a Monochromatic light of wavelength 600 nm falls normally on a single slit of width 0.1 mm. There is a screen 1.0 m in front of the slit. Calculate the following;
  i the angle of the first minimum created on the screen and the middle of the central maximum. [2]

....

....

  ii the distance of the centre of the central maximum to this minimum. [2]

....

....

b Describe the appearance of the central maximum if the slit were illuminated with white light. [2]

....

....

Total   /47

THE COLLEGE OF WEST ANGLIA

Electric current

- An electric current is a flow of charge.
- The size of a current is equal to the rate of flow of charge.
- If charge flows at the rate of 1 coulomb per second (C s⁻¹) then the current is 1 ampere (A).

$$I = \frac{\Delta Q}{\Delta t}$$

$\Delta Q = IC$	$I = 1A$
$\Delta t = Is$	

If a current of 1 A flows, 1 C of charges passes each second.

$$\Delta Q = I \times \Delta t$$

EXAMPLE

Calculate the total charge that passes when a current of 5.0 A flows for 2.0 minutes.

$\Delta Q = I \times \Delta t$
$\Delta Q = 5.0 \times 120$
$\Delta Q = 600\,C$

EXAMINER'S TOP TIP
The majority of the free electrons in a metallic conductor drift towards the positive when a potential difference is applied across it. *Not all of them.*

Metals

In metals the charge carriers are **free electrons**. These free electrons are not held strongly by the positive nucleus of the metal atom and are therefore able to drift from atom to atom. Under normal circumstances, the number of electrons drifting in one direction is equal to the number drifting in the opposite direction. As a result there is no net flow of charge, i.e. there is no current. When a potential difference (see page 48) is applied across a piece of metal, more of the free electrons flow towards the positive terminal and away from the negative terminal. This net flow of charge is an electric current.

Non-metals

In many non-metals there are no (or very few) free charge carriers so current is unable to pass through them. They are non-conductors or insulators.

Gases

Gases do not normally contain charge carriers but, if a high enough voltage is applied between two electrodes, some gas atoms may become ionised allowing current to pass between the plates

Liquids

Currents can pass through liquids if they contain charge carriers called **ions**.
When a voltage is applied between the electrodes, **positive ions** drift towards the negative electrode and **negative ions** drift towards the positive electrode.

Charge carriers

No net flow of charge i.e. no current

More electrons now flow towards the positive side of the metal i.e. there is current

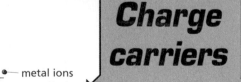
Current passing through a liquid.

Positive ions move towards negative electrode.

Negative ions move towards positive electrode.

EXAMINER'S TOP TIP
Insulators have no free charge carriers.

Drift velocity of electric current

The electric current in a wire is equal to the rate of flow of charge.

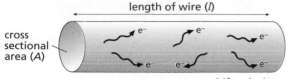

length of wire (l)

cross sectional area (A)

average drift velocity of electrons (v)

In the wire above, there are n free electrons per unit volume. These free electrons each carry a charge of e and are moving with an average drift velocity of v. The volume of the section of wire shown is $A \times l$.

Therefore the number of free electrons in this volume must be nAl.

The total free charge in this volume is therefore $nAle$.

To calculate the time it would take for all these to pass through one point in the wire we use

$$time = \frac{distance}{speed}$$

i.e. $t = \frac{l}{v}$

But

current (I) = $\frac{charge}{time}$

Therefore $I = \dfrac{nAle}{\dfrac{l}{v}}$

$$I = nAve$$

A more general equation for all kinds of charge carriers can be written like this:

$$I = nAvq$$

where q is the charge carried by the charge carrier.

EXAMINER'S TOP TIP

This is an important equation. Make sure you understand what it describes and are able to use it.

EXAMPLE

A copper wire of cross-sectional area 2.5 mm^2 has a current of 1.5 A passing through it. If there are 8.5×10^{28} free electrons in every cubic metre of copper, and the charge on each electron is -1.6×10^{-19} C, calculate the drift velocity of the free electrons in the wire.

$$I = nAvq$$

$$v = \frac{I}{nAq}$$

$$v = \frac{1.5}{8.5 \times 10^{28} \times 2.5 \times 10^{-6} \times 1.6 \times 10^{-19}}$$

$$v = 4.4 \times 10^{-5} \text{ m s}^{-1}$$

Conventional current

When scientists first experimented with electricity, they guessed that the charge carriers in a wire were positively charged and therefore flowed away from the positive terminal and towards the negative. We now know that this is incorrect and that electrons carry a negative charge and therefore flow from the negative terminal to the positive. Nevertheless, it has been agreed to continue to think of current flowing from positive to negative.

This current is called conventional current.

Quick test

1 Calculate the current that flows when 1.0×10^{18} electrons pass a point in a wire in 3 s. ($e = -1.6 \times 10^{-19}$ C).

2 Calculate the number of electrons passing each point in a wire per second when a current of 1.0 A is flowing through it.

3 Explain why metals are good conductors.

4 Explain why polythene is an insulator.

5 The current in a copper wire is 2.7 A. The number of free electrons in each cubic metre of copper is 8.4×10^{28}, the average drift velocity of the electrons is 1.0×10^{-4} m s^{-1} and the charge on an electron is -1.6×10^{-19} C, calculate the cross-sectional area of the wire.

1. 0.05A 2. 6.3 × 10^{18} 3. Some of the electrons in metal atoms are able to drift from atom to atom. 4. There are no free charge carriers in polythene. 5. 2.0 × 10^{-6} m^2 (2.0 mm^2).

45

Resistance and resistivity

- Connecting wires made from copper offer very little opposition to the movement of electrons. They have a <u>low electrical resistance</u>. We measure the resistance in ohms (Ω).

The resistance of a piece of wire (R) at room temperature depends upon

| ● the length of the wire (l) | ● the cross-sectional area of the wire (A) | ● the <u>resistivity</u> (ρ) of material from which the wire is made. |

These quantities are related by this equation:

$$R = \frac{\rho l}{A}$$

EXAMPLE

Calculate the resistance of a copper wire 50 cm long and with a cross-sectional area of 2.0 mm². The resistivity of copper is $1.7 \times 10^{-8}\ \Omega$ m.

Using $\quad R = \frac{\rho l}{A}$

$$R = \frac{1.7 \times 10^{-8} \times 0.50}{2.0 \times 10^{-6}}$$

$$R = 4.3 \times 10^{-3}\ \Omega$$

The <u>conductivity</u> of a material (σ) is equal to the reciprocal of its resistivity.

$$\sigma = \frac{1}{\rho}$$

The conductivity of copper is therefore

$$\frac{1}{1.7 \times 10^{-8}} = 5.9 \times 10^{7}\ \Omega^{-1}\ \text{m}^{-1}$$

$$6 = \frac{1}{1.7} \times 10^{-8} = 5.9 \times 10^{7}\ \Omega^{-1}\ \text{m}^{-1}$$

Resistance and temperature

As electrons drift through a metal they collide with the lattice structure, transferring energy to it. If the metal is warmed the atoms within the lattice vibrate more vigorously and the frequency of the electron collisions increases, i.e. the resistance of a metallic conductor increases with temperature.

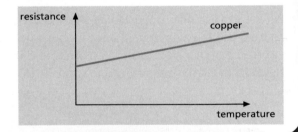

Semiconductors

Not all materials show an increase in resistance as temperature rises. Silicon and germanium are semiconductors. They have some free charge carriers, but far fewer than those found in most metals. If, however, they are warmed, more charge carriers become free and the movement of charge through the material is easier. The resistance of a semiconductor decreases as its temperature increases.

The resistance of a **thermistor** is very sensitive to changes in temperature and they are therefore often used in temperature sensing circuits (see potential dividers page 51).

The number of free charge carriers in a **light dependent resistor** (**LDR**) is affected by the intensity of the light falling upon it, i.e. the more intense the light, the more charge carriers become available for conduction and so the lower the resistance. LDRs are often used in light-sensitive circuits such as automatic lighting controls (see potential dividers page 51).

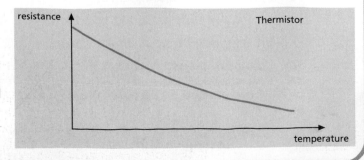

Measuring resistance

The resistance (R) of a component in a circuit is defined as being equal to the ratio of the voltage (V) across the component to the current (I) that flows through it.

$$R = \frac{V}{I}$$

If a voltage of 1 V is applied across the ends of a component and this causes a current of 1 A to flow, the resistance of the component is 1 Ω.

The circuit below shows how the resistance of a component can be measured.

The switch, S, is closed and the current in the wire (I) and the p.d. across it (V) are noted.

The value of the variable resistor is altered and the new values for I and V are noted.

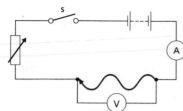

The experiment is repeated until at least six pairs of readings are obtained.

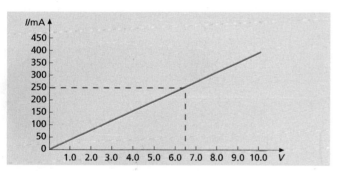

A graph of I against V is then plotted.

The graph shows that, for a metallic conductor such as a piece of wire, the current that flows is directly proportional to the p.d. applied across its ends.

Components that show this behaviour are known as **ohmic conductors** as they obey **Ohm's law**.

Ohm's law states that the current in a metallic conductor is directly proportional to the potential difference across its ends, provided that the temperature remains constant.

The resistance of the wire can be found by taking values from the graph and using the equation

$$R = \frac{V}{I}$$

In this case the resistance of the wire is

$$\frac{V}{I} = \frac{6.5}{250 \times 10^{-3}} = 26\ \Omega$$

Most common components are not ohmic conductors. As the current flowing through a bulb increases, temperature of the filament wire increases and so too does its resistance.

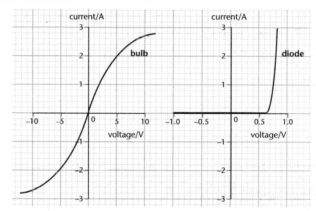

Diodes are components that exhibit a different resistance to current depending upon the direction in which it is flowing, i.e. in one direction the resistance is low, whilst in the opposite direction resistance is high. Diodes are often used in rectifiers to change alternating currents into direct currents.

Quick test

1 The resistivity of lead is $2.1 \times 10^{-7}\ \Omega$ m. Calculate the resistance of a piece of lead 150 cm long and with a cross-sectional area of 4.0 mm².

2 Calculate the conductivity of lead.

3 Most thermistors have a negative temperature coefficient (NTC). Explain what this means.

4 A current of 200 mA flows when a p.d. of 5 V is applied across a wire. Calculate the resistance of the wire.

5 What is an ohmic conductor? Give one example of an ohmic conductor.

6 Sketch the current–voltage graph for a thermistor (NTC). Explain the shape of the graph you have drawn.

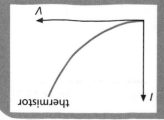

1. 0.079 Ω 2. $4.8 \times 10^6\ \Omega^{-1}\ m^{-1}$ 3. As the temperature of the thermistor increases, its resistance decreases. 4. 25 Ω 5. An Ohmic conductor is one which obeys Ohm's law, i.e. the current in the conductor is directly proportional to the p.d. across its ends, provided that the temperature remains constant. A metal wire at constant temperature is an example of an ohmic conductor. 6. As the current flowing through the thermistor increases, its temperature increases and so its resistance decreases.

Electrical potential difference

- Energy is stored in cells and batteries as chemical energy.
- When charges flow through them they receive some of this energy.
- The energy received is called electrical energy.
- The energy transferred from chemical to electrical per unit charge that passes through the cell/battery is known as the <u>electromotive force</u> (e.m.f.) of the supply and is measured in volts.

A voltmeter connected across the terminals of a cell as shown here is measuring the e.m.f. of the cell. The meter is reading 1.5 V. This indicates that each coulomb of charge flowing through the cell is receiving 1.5 J of energy. A reading of 1.8 V would indicate that 1.8 J of energy was being given to each coulomb of charge, etc.

This can be summarised by this equation:

e.m.f. of source $= \dfrac{\text{energy transfer}}{\text{charge}}$

$$\boxed{or \quad V = \dfrac{W}{Q}}$$

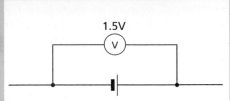

From the above we can see that <u>1 volt = 1 joule per coulomb</u>.

EXAMPLE

Calculate the e.m.f. of a battery that gives 400 J of its chemical energy to 200 C of charge that pass through it.

$V = \dfrac{W}{Q} = \dfrac{400\,J}{200\,C} = 2.0\,V$

Potential difference

As charges flow around a circuit, the energy they receive from the cell or battery is transformed into other forms of energy by the components in the circuit.

If a voltmeter is connected in parallel with any of the components, it will measure the potential difference (p.d.) across that component. The potential difference across a component indicates how much energy is transformed as charges flow through it.
If the p.d. across a component is 1 V, then 1 J of electrical energy is converted into 1 J of a different form of energy when 1 C of charge flows through it.

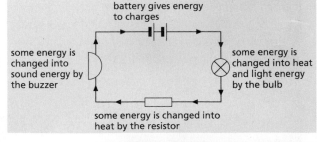

- All the energy given to the charges by the battery is converted into other forms of energy. Therefore, for a series circuit like this where there is only one path for the charges to follow, $V_b = V_1 + V_2 + V_3$
- The energy dissipated in the connecting wires is assumed to be negligible compared with that dissipated by the other components.
- The relationship between the amount of energy transformed (W), the charge passing through the component (Q) and the p.d. across it (V) is given by this equation: $W = Q \times V$

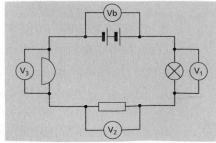

EXAMPLE

Calculate the total energy dissipated in the bulb above, where $V_1 = 1.5$ V, when 20 C of charge pass through it.

$$W = Q \times V = 20\,C \times 1.5\,V = 30\,J$$

Energy is dissipated in a circuit because work is being done in moving charge through the various components.
   If the p.d. across a component is 1 V, then 1 J of work must be done to move 1 C of charge through it.

Electromotive force and internal resistance

In each of the previous circuits, we have assumed that all the energy received by the charges from a cell or battery is dissipated in the external part of the circuit, i.e. outside the cell/battery. In practice, this is not always true. Energy may be needed to move charge through the cell/battery.

i.e. $E = V + v$

where E is the e.m.f. of the cell (the energy given to each coulomb of charge that passes through it)

V is the potential difference across the terminals of cell/battery (the energy dissipated by each coulomb of charge in the external part of the circuit).

v is often called the *lost volts* and is the energy dissipated by each coulomb of charge within the cell/battery.

Work is done by the charges within the electrical supply because the cell/battery has internal resistance (r).

Since energy is conserved E = sum of the p.d.s in the circuit i.e.

$$E = IR + Ir \quad \text{or} \quad E = I(R + r)$$

(R is the resistance of the external part of the circuit.)

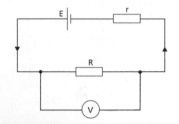

EXAMPLE

Calculate the internal resistance of a 1.5 V cell which produces a current of 0.10 A when connected to an external resistance of 14 Ω.

$$E = I(R + r)$$
$$1.5 = 0.10(14 + r)$$
$$r = 1.0 \ \Omega$$

- As the internal resistance of a cell increases, less energy is available to do work in the external part of the circuit.
- When a short circuit occurs, the internal resistance of the supply may limit the current that flows.

EXAMINER'S TOP TIP
The internal resistance of a cell may limit the maximum current that can flow in a circuit. This may be an advantage in some circuits and a disadvantage in others.

Measuring e.m.f. and internal resistance

With the switch closed, the p.d. across the cell (V) and the current (I) flowing around the circuit are noted.

The value of the variable resistor is altered and a new pair of readings for V and I are noted.

This procedure is repeated until a minimum of six pairs of readings is obtained.

From $E = IR + Ir$

we can write $E = V + Ir$

or $V = -Ir + E$

This equation is of the form

$y = mx + c$

Plotting a graph of V against I will therefore produce a straight-line graph whose gradient is equal to $-r$ and whose intercept on the y-axis is equal to the e.m.f. of the cell/battery.

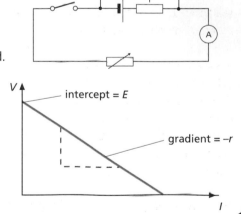

Quick test

1 Calculate the electrical energy transformed into sound energy when 10 C of charge pass through a buzzer which has a p.d. of 2.0 V across it.

2 Describe the relationship between the e.m.f. of a cell and the p.ds across all the external components in a series circuit, assuming the internal resistance of the cell is negligible.

3 A current of 0.25 A flows when a battery of e.m.f. 12 V is connected to a circuit that has a total external resistance of 46 Ω. Calculate the internal resistance of the battery.

Potential dividers

- If several resistors are connected in series the total resistance of the network (R_t) can be calculated using this formula:

$$R_t = R_1 + R_2 + R_3 + \dots$$

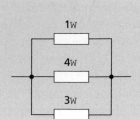

- The total resistance of this network of resistances is therefore $5\,\Omega + 10\,\Omega + 20\,\Omega = 35\,\Omega$

- If several resistors are connected in parallel, the total resistance of the network can be calculated using this formula.

$$\frac{1}{R_t} = \frac{1}{R_1} + \frac{1}{R_2} + \frac{1}{R_3} + \dots$$

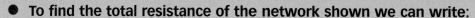

- To find the total resistance of the network shown we can write:

$$\frac{1}{R_t} = \frac{1}{1} + \frac{1}{4} + \frac{1}{3} = \frac{12 + 3 + 4}{12}$$

$$\frac{1}{R_t} = \frac{19}{12}$$

$$R_t = 0.63\,\Omega$$

EXAMINER'S TOP TIP
Questions on working out the total resistance of resistors connected in series and in parallel are very common.

The potential divider

If a voltage is applied across several resistors in series, the p.d. is shared across all of them according to their relative resistances.

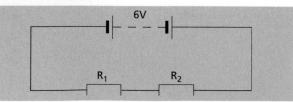

If R_1 and R_2 have the same value, the p.d. across each will, in this case, be 3 V. If R_1 has twice the resistance of R_2, it will have twice the p.d. across it, i.e. there will be a p.d. of 4 V across R_1 and a p.d. of 2 V across R_2.

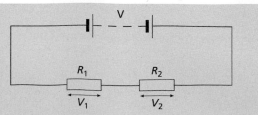

In general, the p.d. across R_1 is given by

$$V_1 = \left(\frac{R_1}{R_1 + R_2}\right) \times V$$

Similarly, the p.d. across R_2 is given by

$$V_2 = \left(\frac{R_2}{R_1 + R_2}\right) \times V$$

EXAMPLE
Calculate the potential difference across the 10 Ω resistor in the circuit shown below.

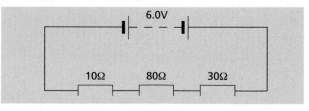

$$V_{10} = \left(\frac{10}{10 + 80 + 30}\right) \times 6.0 = 0.5\,V$$

Altering the resistances in the network alters the p.d. across each of the resistors.

EXAMINER'S TOP TIP
Make sure you can use this equation.

Uses of potential dividers

Potential dividers are often used in sensing circuits. The circuit below uses a thermistor as part of its potential divider and is a **temperature control circuit**.

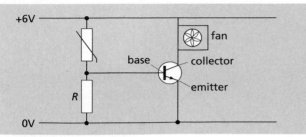

An increase in temperature causes the resistance of the thermistor to decrease and the p.d. across the resistor R to increase. This increase in voltage switches the transistor* on, which allows current to flow through the fan.

---

* Transistors are semiconductor devices that can be used as switches. If the potential difference between the base and the emitter is high, current is allowed to flow between the emitter and the collector, i.e. the transistor is switched on. If the p.d. between the base and the emitter is low, current is unable to flow between the emitter and the collector, i.e. the transistor is switched off.

EXAMINER'S TOP TIP
Make sure you can draw several circuits that show how thermistors and LDRs can be used in potential dividers.

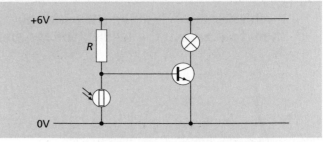

This circuit uses an LDR as part of its potential divider and is a **light-sensing circuit**. It may be used for automatic lighting control. As the intensity of the light falling on the LDR decreases, its resistance increases. The potential difference across the LDR increases, and eventually switches on the transistor, resulting in the bulb glowing. If the fixed resistor R in each of the circuits illustrated is replaced with a variable resistor, the temperature at which the fan is turned on and the level of the light intensity at which the bulb is turned on can be adjusted.

In some circuits, the resistors that form the potential divider are replaced by a uniform resistance with a sliding contact. As the contact is moved, the values of R_1 and R_2 are continuously variable. The p.ds across each resistor can therefore have all values between 0 and 12 V.

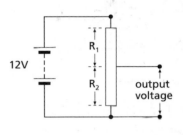

Quick test

1 Calculate the resistance between A and B for the following network of resistors.

a
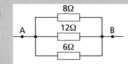

b
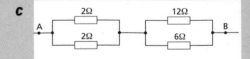

c
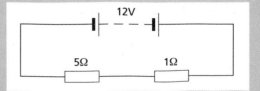

2 a Calculate the p.d. across the 5 Ω resistor and the 1 Ω resistor in the circuit (right).

b Calculate the p.d. across each of the resistors if the 1 Ω resistor is replaced with a 3 Ω resistor.

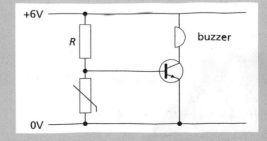

3 Look carefully at the circuit (right). What is this circuit being used for?

1.a) 3.0 kΩ   b) 2.7 Ω   c) 5.0 Ω   2.a) 10 V and 2 V respectively   b) 7.5 V across the 5 Ω resistor and 4.5 V across the 3 Ω resistor
3. The circuit is being used to monitor falling temperatures, i.e. the buzzer sounds when the temperature falls to a particular value.

51

Kirchhoff's laws

Kirchhoff's laws describe the behaviour of currents and voltages in circuits. We use them to calculate the applied voltage and the current flowing at any point in a circuit.

Kirchhoff's first law

When charges flow into a junction the same charge must also flow out of the junction. It is not possible for charge to disappear or build up in the junction. This **conyorvation of charge** is summarised by Kirchhoff's first law.

The algebraic sum of the currents flowing into a circuit junction must be equal to the algebraic sum of the currents flowing out of that junction.

Applying Kirchhoff's first law to this junction we can write:

$$I_1 + I_5 + I_6 = I_2 + I_3 + I_4$$

EXAMPLE

Calculate the current X.

$$4\,A + 5\,A = X + 2\,A + 1\,A$$
$$X = 6\,A$$

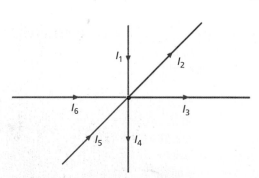

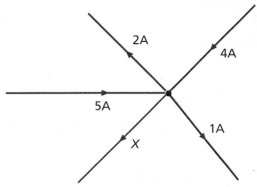

Kirchhoff's second law

When charges pass through cells/batteries they gain electrical energy. As they travel around a complete circuit or loop, all this energy is dissipated. This **conservation of energy** is summarised by Kirchhoff's second law.

The algebraic sum of the e.m.fs in any closed loop is equal to the algebraic sum of the p.ds around that loop.

$$\Sigma E = \Sigma IR$$

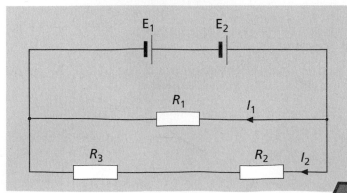

Applying Kirchhoff's second law to this circuit we can write:

For the upper loop $E_1 + E_2 = IR_1$

For the outside loop $E_1 + E_2 = I_2R_2 + I_2R_3$.

EXAMINER'S TOP TIP
It is easy to lose your way with these calculations. Set out your workings in a neat, logical way so that it is easy to double-check them.

Example using Kirchhoff's second law

Apply Kirchhoff's First Law at junction B

$I_1 + I_2 = I_3$ (1)

Applying Kirchhoff's Second Law to the loop ABEF

$12 = 6I_1 + 12I_3$ (2)

Applying Kirchhoff's Second Law to the loop BCDE

$6 = 6I_2 + 12I_3$ (3)

Adding equations (2) and (3) we get

$18 = 6I_1 + 6I_2 + 24I_3$

From equation (1) therefore

$18 = 6I_3 + 24I_3 = 30I_3$

therefore $I_3 = \dfrac{18}{30} = 1.6$ A

Substituting in equation (2):

$12 = 6I_1 + 12 \times 0.6$

$I_1 = \dfrac{4.8}{6} = 0.8$ A

Substituting in equation (1):

$I_1 + I_2 = I_3$

$0.8\,A + I_2 = 0.6$

$I_2 = -0.2$ A   – The minus sign indicates that I_2 flows in the opposite direction to that shown on the diagram

EXAMPLE Using Kirchhoff's Laws calculate the currents I_1, I_2 and I_3

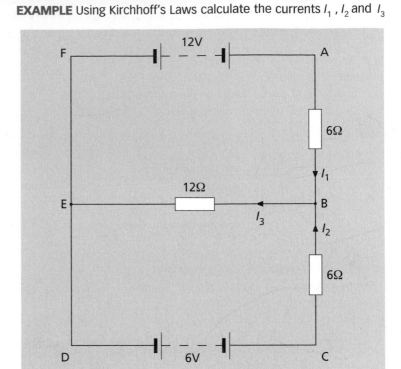

Quick test

1 Explain why Kirchhoff's first law might also be called 'the law of conservation of charge'.

2 Explain how Kirchhoff's second law is in agreement with the law of conservation of energy.

3 Using Kirchhoff's laws, calculate the values of I_1 and E_1 in the circuit shown below.

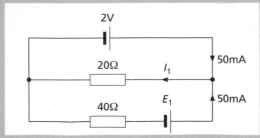

4 Using Kirchhoff's laws, calculate the values of I_1 and R in the circuit shown below.

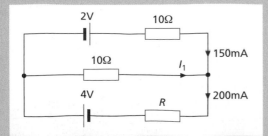

1. Kirchhoff's first law can be stated in terms of charge rather than current, i.e. the charges entering a junction must be equal to those leaving the junction. 2. Kirchhoff's second law states that the energy received from the supply is equal to the energy delivered to the different parts of the circuit by the current. 3. $I_1 = 100$ mA, $E_1 = 4.0$ V   4. $I_1 = 50$ mA, $R = 18$ Ω

53

Electrical energy and power

Energy transfer

The potential difference between two points in a circuit is a measure of the electrical energy transferred to, or from, 1 C of charge as it moves from one point to the other. The potential difference between two points can be measured using a voltmeter.

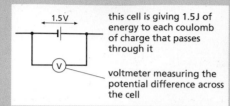

this cell is giving 1.5 J of energy to each coulomb of charge that passes through it

voltmeter measuring the potential difference across the cell

This voltmeter is measuring the energy given to each coulomb of charge as it passes through the cell.

This voltmeter is measuring the energy transferred into heat and light from each coulomb of charge as it passes through the bulb.

We can express these energy transfers using this equation:

$$\text{potential difference} = \frac{\text{energy transferred}}{\text{charge}}$$

or

$$V = \frac{W}{q}$$

Rearranging this equation we get:

$$W = Vq$$

but $\quad q = I \times t$

therefore $\quad \boxed{W = VIt}$

Examples of energy power

EXAMPLE

Calculate the energy transferred by a bulb when a current of 200 mA passes through it for 100 s. The potential difference across the bulb is 6 V.

$W = VIt = 6 \times 0.2 \times 100 = 120$ J

EXAMPLE

Calculate the energy transferred from a 1.5 V cell when a current of 0.50 A passes through it for 5 minutes.

$W = VIt = 1.5 \times 0.50 \times 5 \times 60 = 225$ J

Using $V = IR$, we can write the equation $W = VIt$ in several different forms:

$$\boxed{W = RI^2t} \quad \text{and} \quad \boxed{W = \frac{V^2}{R} t}$$

EXAMPLE

Calculate the total energy transferred when a current of 0.20 A passes through a 100 Ω resistor for 50 s.

$W = RI^2t = 100 \times (0.20)^2 \times 50 = 200$ J

EXAMPLE

Calculate the amount of electrical energy converted into heat and light in 20 s by a bulb that has a p.d. of 2.0 V across it and a resistance of 200 Ω.

$W = V^2 \dfrac{t}{R} = (2.0)^2 \times \dfrac{20}{200} = 0.4$ J

Electrical power

The rate at which a component transfers energy is its **power** (P) and is measured in **watts** (W). A component which has a power of 1 W is transferring energy at the rate of 1 joule per second.

This bulb has a power rating of 40 W. It is changing 40 J of electrical energy into 40 J of heat and light energy every second.

From the three energy equations above we can therefore write three equations that we can use to determine the power of a component:

$$\boxed{\begin{array}{l} P = VI \\ P = I^2R \\ P = \dfrac{V^2}{R} \end{array}}$$

EXAMPLE

Calculate the power of a bulb which has a resistance of 180 Ω when a current of 0.6 A is passing through it.

$P = I^2R = (0.6)^2 \times 180 = 64.8$ W

EXAMPLE

Calculate the rate at which electrical energy is changed into heat energy by a heater that has a resistance of 1 kΩ when a p.d. of 200 V is applied across it.

$P = \dfrac{V^2}{R} = \dfrac{200^2}{1000} = 40$ W

Using electrical energy in the home

When electrical appliances such as televisions and dishwashers are turned on, they convert large amounts of electrical energy into other forms. The joule is too small a unit of energy to conveniently measure these transfers. Electricity boards therefore use a different energy measurement.
It is called a __unit__ or a __kilowatt-hour__ (__kWh__).

A device which has a power rating of 1000 W or 1 kW will, if turned on for 1 h, transform 1 kWh of electrical energy.
We can calculate the energy used by such appliances using this equation:

EXAMPLE
A 3 kW electric fire is turned on for 4 h. How much electrical energy does the fire convert into heat energy?

energy used = 3 kW × 4 h = 12 kWh or 12 units

energy used = power of appliance in kilowatts × number of hours

Transmitting electrical energy

The electrical energy we use in the home is usually produced by power stations and then transmitted to us along a network of cables and wires called the
__National Grid__.

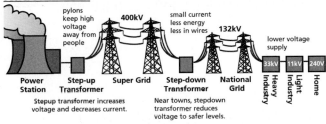

Power Station	Step-up Transformer	Super Grid	Step-down Transformer	National Grid	Heavy Industry	Light Industry	Home

Stepup transformer increases voltage and decreases current.

Near towns, stepdown transformer reduces voltage to safer levels.

As this energy travels through the cables, some of it is wasted. It is changed into heat and warms the cables. The amount of energy lost per second is given by the equation $P = I^2R$ where R is the resistance of the cables.

To reduce this energy loss, a step-up transformer is used so that the energy is transmitted with a high voltage and a much smaller current.

EXAMPLE
Electrical energy is to be transmitted along a power cable of resistance 1.0 Ω at 100 V and 100 A, or at 10 000 V and 1.0 A. Calculate the rate of energy loss in the cables for each set of conditions.
rate of energy loss with the first set of conditions = I^2R
= $(100)^2 × 1.0$ = 10 000 W
rate of energy loss with the second set of conditions = I^2R
= $(1.0)^2 × 1.0$ = 1.0 W
The calculations clearly show that transmitting electrical energy at high voltages and low current greatly reduces energy loss in the cables.

Quick test

1 Calculate the total energy transfer in each of these.

  a A current of 0.5 A flowing through a 200 Ω resistor for 60 s.

  b A p.d. of 10 V applied across a bulb whose resistance is 200 Ω for 10 s.

  c A current of 100 mA flowing for 1 minute through a resistor which has a p.d. of 20 V applied across it.

2 Calculate the energy transferred to or from flowing charges in each of these situations.

  a 10 C of charge flowing through a 6 V battery.

  b 15 C of charge flowing through a bulb which has a p.d. of 3 V across it.

3 Calculate the energy dissipated as heat when a current of 0.5 A flows through a 100 Ω resistor for 1 minute.

4 Calculate the number of units of electrical energy that are converted into heat when an electric fire rated at 2 kW is turned on for 3 h.

Exam-style questions
Use the questions to test your progress. Check your answers on pages 92–95.

1 a Explain why metals are good conductors of electricity. [2]

....
....

b Explain how electrical current is able to pass through some liquids. [2]

....
....

c A copper wire of cross-sectional area 2.0 mm² has a current of 1.0 A passing through it. If there are 8.5×10^{28} free electrons in every cubic metre of copper, and the charge on each electron is 1.6×10^{-19} C, calculate the drift velocity of the free electrons in the wire. [4]

....
....
....
....

d Explain why the average drift velocity of the electrons would decrease if the copper wire were to be warmed. [2]

....
....

2 a Explain the difference between resistance and resistivity. [2]

....
....

b Calculate the resistance of a piece of copper wire that is 75 cm long and has a diameter of 2.0 mm. The resistivity of copper is 1.55×10^{-8} Ω m. [3]

....
....
....

c The heating element of an electric kettle is made from nichrome wire that has a diameter of 0.05 mm and a resistance, at room temperature, of 30 Ω. If the resistivity of nichrome is 1.1×10^{-6} Ω m, calculate the length of the wire. [3]

....
....
....

3 A large torch has four identical 1.5 V cells connected in series as its power supply. When switched on, the potential difference across the bulb is 5.4 V and it dissipates energy at the rate of 1.0 W. Calculate these.
a the current flowing through the bulb [2]

....
....

b the internal resistance of each of the cells [2]

....
....

c the energy dissipated in each of the cells in 5 minutes [2]

....
....

4 The diagram opposite shows a 12 V battery with negligible internal resistance connected across a network of resistors.
Calculate.
a the effective resistance between A and B [2]

....
....

b the effective resistance between B and C [2]

....

....

c the current through each of the 2 Ω resistors [2]

....

....

d the current through the 4 Ω resistor [2]

....

....

5 The diagram shows a simple potential divider.
 a Calculate the potential difference across
  i the thermistor   **ii** the resistor R [4]

....

....

....

....

....

 b Explain what happens to each of these potential differences if the temperature of the
  thermistor increases. [2]

....

....

 c Name one device which might contain a circuit similar to that drawn above. [1]

....

6 a State Kirchhoff's laws. [2]

....

....

 b Using Kirchhoff's laws, calculate the currents I_1, I_2 and I_3 shown in the circuit below. [6]

....

....

....

....

....

....

....

7 a What is an ohmic conductor? Give one example of an ohmic conductor. [2]

....

....

 b Sketch a current–voltage graph for an ohmic conductor. [2]

....

....

 c Name one non-ohmic conductor. [1]

....

 d Sketch the current–voltage graph for the non-ohmic conductor named in part c). [2]

....

....

Total   /54

The photoelectric effect and wave–particle duality

- If electromagnetic radiation strikes the surface of a metal, it can cause electrons to be emitted. This phenomenon is called <u>photoelectric emission</u> or the <u>photoelectric effect</u>. Energy is being given to electrons by the radiation and is being used to enable them to escape the metal.

The photoelectric effect

To demonstrate the photoelectric effect a zinc plate is placed on the cap of a gold-leaf electroscope. The zinc plate and the cap are charged negatively. This is confirmed by the deflection of the gold leaf. If photoelectric emission takes place, the plate and electroscope will **lose negative charge** and the gold leaf will fall.

When **visible light** is shone on the zinc plate, photoelectric emission does not take place even when the intensity (brightness) of the light is increased.

However, when **ultraviolet radiation** is shone on the plate, the leaf falls, showing that photoelectrons are being emitted. Even with **low-intensity** ultraviolet radiation the leaf falls, but it does so more slowly.

Further similar experiments with other types of radiation lead to these conclusions.

- Photoelectric emission takes place only if the frequency of the incident radiation is greater than a certain minimum value called the **threshold frequency (f_0)**.
- Different metals have different threshold frequencies.
- The intensity of an incident radiation does not affect whether photoemission takes place but it does affect the rate of emission of photoelectrons, i.e. if the radiation has a frequency above the threshold frequency, increasing its intensity will increase the rate at which the photoelectrons are emitted.

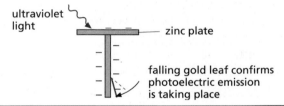

zinc plate — negatively charged gold-leaf electroscope

gold-leaf is deflected confirming that the electroscope is charged

ultraviolet light — zinc plate

falling gold leaf confirms photoelectric emission is taking place

The energies of photoelectrons

An investigation

If radiation with a frequency above the threshold frequency is shone onto the metal electrode A, photoelectrons are emitted. Some of these will reach B and this flow of charge (current) will be detected on the microammeter. If a p.d. is applied between electrode A and electrode B, with B negative with respect to A, electrons travelling from A to B will do work against the electric field. This work will be done at the expense of the kinetic energy of the photoelectrons. If the p.d. is gradually increased, the current registered by the microammeter decreases. At the point at which the current registered becomes zero, the work done against the field must be equal to the maximum kinetic energy of photoelectrons. This voltage is known as the **stopping potential (V_s)**.

If the experiment is repeated with radiations of different frequencies similar results are obtained, but the stopping potential increases as the frequency of the radiation increases.

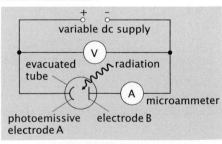

variable dc supply

evacuated tube

radiation

photoemissive electrode A

electrode B

microammeter

Conclusions drawn from experiment

- The photoelectrons are emitted with a range of energies from zero to a maximum value.
- If the frequency of the radiation is increased, so too is the maximum kinetic energy of the photoelectrons.
- For a fixed frequency, the maximum kinetic energy of the photoelectrons is unaffected by the intensity of the radiation.

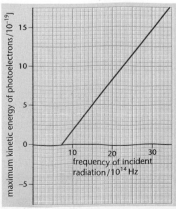

The quantum theory

The wave model of light explains phenomena such as interference and diffraction, but it could not explain how light interacts with matter, that is, electrons. The wave model predicts that whether a photoelectron is emitted or not should depend on the intensity of the radiation and not its frequency. Similarly, the maximum kinetic energy of the photoelectrons should depend only on the intensity of the incident radiation. Neither of these were shown to be correct by the previous experiment.

- To explain these findings, another model was needed. It was called the **quantum theory**.
- The quantum theory suggests that electromagnetic radiation consists of a stream of **quanta** or pockets of energy
- The energy carried in each **quantum** is given by
$$E = hf$$
where f is the frequency of the radiation and h is a constant called the **Planck constant** (6.63×10^{-34} J s).

Quanta and the photoelectric effect

- A pocket of light energy is called a **photon**.
- When a photon interacts with an electron it transfers all its energy. The energy transfer is always one to one, it can never be shared.
- If the frequency of the radiation is less than the threshold frequency, an electron will not receive enough energy for it to escape the surface of the metal.
- If the frequency is equal to the threshold frequency, an electron will receive just enough energy to escape. The minimum amount of energy that an electron needs to escape from a metal is called the **work function (ϕ)**. The work function of a metal is usually expressed in electron volts where 1ev = 1.6×10^{-19} J.

Different metals have different work functions.

- It follows that the relationship between the work function (ϕ) and the threshold frequency (f_0) is given by the equation
$$\phi = hf_0$$

- If the frequency of the incident radiation is greater than the threshold frequency, electrons will escape and have surplus energy in the form of kinetic energy.

If a photon interacts with an electron within the metal, some of the electron's energy may be lost because of collisions with ions as it moves towards the surface. It is therefore emitted with less kinetic energy. This is why photoelectrons are emitted with a range of kinetic energies.

Electrons that are emitted without losing energy in interactions will have the maximum kinetic energy for that frequency of radiation. This is described by the equation

$$hf = \phi + \tfrac{1}{2} mv_{max}^2$$

Einstein's photoelectric equation

Wave-particle duality

If waves can exhibit properties that we would normally associate with particles (photons), perhaps particles can exhibit some of the properties we normally associate with waves. The diagram opposite confirms that this is the case. In 1924 the French physicist Louis de Broglie suggested that all moving particles possessed some wave –like properties and that the relationship between the momentum of the particle (p) and its wavelength (λ) was described by the equation $\lambda = \dfrac{h}{p}$

Diffraction pattern formed by a beam of electrons passing through graphite

EXAMPLE. Calculate the de Broglie wavelength of an electron moving with a speed of 1.5×10^7 ms^{-1}.
($h = 6.6 \times 10^{-34}$ Js and $m_e = 9.1 \times 10^{-31}$ kg)
$$\lambda = \frac{h}{p} = \frac{6.6 \times 10^{-34}}{9.1 \times 10^{-31} \times 1.5 \times 10^7} = 4.8 \times 10^{-11} \text{ m}$$

Quick test

1 **The threshold frequency of a metal is 9.0×10^{14} Hz. Calculate the work function of this metal in joules and electron volts.**

2 **Calculate the maximum kinetic energy of photoelectrons emitted from a metal with a work function of 4.3 eV when the incident radiation has a frequency of 2.0×10^{15} Hz.**

3 **Calculate the de Broglie wavelength of an electron travelling at 5.0×10^7 m s^{-1}.**

1. 5.9×10^{-19} J and 3.7 eV 2. 6.3×10^{-19} J or 3.9 eV 3. 1.5×10^{-11} m

Spectra – Emission spectra

Line spectra

- Electrons in atoms occupy **energy levels**. They cannot exist between these levels.
- Normally an electron will occupy the lowest energy level available (**ground state**) but, if it absorbs energy, it may be able to **jump** to a higher energy level (**excited**).
- After a short time this electron will return to a lower energy level. To do so, it must lose energy equal to the energy difference between the two energy levels. It does this by emitting a **photon** of electromagnetic radiation.
- The frequency of the photon emitted depends upon the difference in energy levels.
- Atoms of an element have unique differences in electron energy levels. The colours of the light they emit, i.e. their **spectra**, are characteristic of that element and can be used like a fingerprint to identify it.
- For most elements there is the possibility for electrons to make more than just one kind of energy jump or **transition**. The resulting **emission spectra** are seen as a series of coloured lines like that shown to the right.
- The study of spectra is called **spectroscopy**.

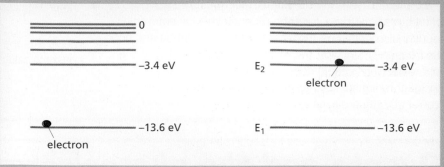

Electron energy levels in a hydrogen atom with the electron occupying the lowest energy level.

The electron has absorbed some energy and made a transition to a higher energy level.

The emission spectrum of cadmium.

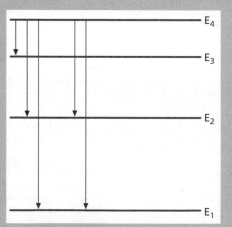

There are several transitions that electrons can make within an atom. So photons with different frequencies are emitted. These different frequencies of light form an emission spectrum.

Band and continuous spectra

Line spectra are produced by individual, isolated atoms – for example, the atoms of a gas. **Band spectra** are produced by molecules, or atoms, that are closer together. The electrons of neighbouring atoms interact because of their proximity, resulting in a broadening of the electron energy levels. As a result, groups of spectral lines or bands are seen rather than individual lines. If there are a large number of atoms and they are very close together, for example in a solid, a very broad spread of wavelengths may be emitted. This is seen as a **continuous spectrum**.

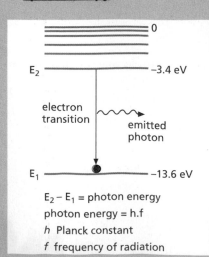

$E_2 - E_1$ = photon energy

photon energy = h.f

h Planck constant

f frequency of radiation

The electron emits a photon of energy as it returns to a lower energy level.

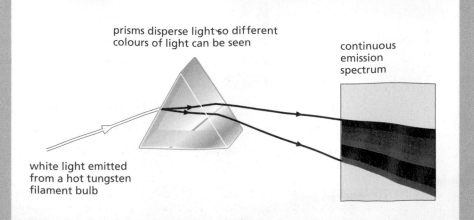

prisms disperse light so different colours of light can be seen

continuous emission spectrum

white light emitted from a hot tungsten filament bulb

Distribution of wavelengths

Distribution within an emission spectrum

If a piece of iron is heated, some of the electrons of its atoms will make transitions which will result in the emission of radiation with wavelengths corresponding to the infra-red part of the electromagnetic spectrum. As the temperature of the iron increases, higher energy photons will be emitted and the spectrum will contain radiation of shorter wavelengths (larger frequencies), i.e. the colour of the iron changes from dull red to bright red, then orange, yellow and then, finally, white. The diagram below shows how the distribution of the wavelengths changes as the temperature of the iron changes.

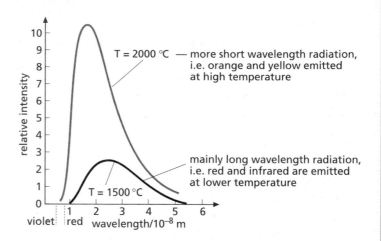

T = 2000 °C — more short wavelength radiation, i.e. orange and yellow emitted at high temperature

mainly long wavelength radiation, i.e. red and infrared are emitted at lower temperature

T = 1500 °C

Absorption spectra

As white light passes through a gas, electrons within the atoms of the gas may absorb energy in order to make transitions to higher energy levels. The wavelengths of the light absorbed will match exactly the energy required to make a particular upward transition. Observation of the spectrum of the white light will show dark lines where these wavelengths of light are missing. This type of spectrum is called an **absorption spectrum**.

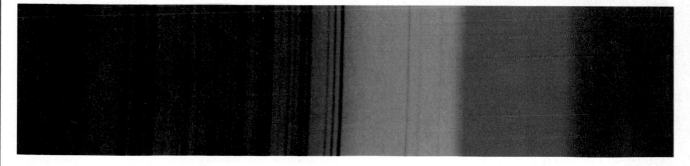

In an absorption spectrum, the lines are characteristic of the atoms in the gas.
Absorption spectra are useful to astronomers. Analysis of the light from a star may reveal details about the temperature of the surface of the star and the elements in the gas surrounding the star.

Quick test

1 What is a photon?

2 How is the energy contained in a photon related to the frequency of the radiation?

3 What is an upward electron transition?

4 What must an electron do to make a downward transition?

5 Why can emission line spectra be used to identify atoms?

6 What kind of information may be discovered by observing the spectra of stars?

1. A small packet of light energy or quantum of radiation. 2. The higher the frequency, the greater the energy carried by the photon. 3. An electron has absorbed energy and moved to a higher energy level. 4. Give out energy. 5. The emission lines correspond to the transitions made by electrons within an atom. These transitions, the energies of the photon emitted and, therefore, the frequencies of the lines seen in a spectrum are unique to that particular atom. 6. The temperature of the stars and the elements of atoms in the gases surrounding them

The electromagnetic spectrum

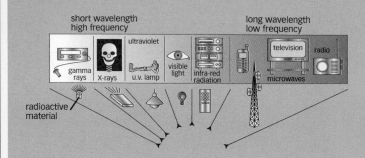

short wavelength
high frequency

long wavelength
low frequency

gamma rays
X-rays
ultraviolet
u.v. lamp
visible light
infra-red radiation
television
microwaves
radio

radioactive material

This is a family of waves, including visible light, with a large number of common properties.

- They are all able to travel through a vacuum.
- They all travel at the same speed through a vacuum (3×10^8 m s^{-1}).

- They are all transverse waves consisting of magnetic and electric fields oscillating at right angles to each other.
- They all transfer energy. This energy is carried in small packets or quanta called photons. The higher the frequency of a particular radiation, the greater the energy contained in each photon, i.e. a photon of ultraviolet light contains more energy than a photon of infra-red.
- They can all be reflected, refracted, diffracted and create interference patterns.
- Some of the properties of these waves change as the wavelength and frequency change. The spectrum is therefore divided into seven groups.

Radio waves

- Radio waves are used for **communicating over large distances**.
- Short wavelength radio waves are used for television broadcasting and FM radio.
- Longer wavelength radio waves are used for traditional AM radio.
  Typical values
- Long wave $\lambda = 1 \times 10^4$ m, $f = 3 \times 10^4$ Hz
- Medium wave $\lambda = 1 \times 10^2$ m, $f = 3 \times 10^6$ Hz
- Short wave $\lambda = 1 \times 10^0$ m, $f = 3 \times 10^8$ Hz

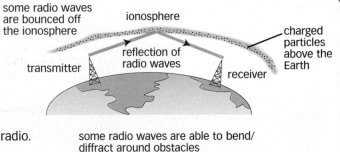

some radio waves are bounced off the ionosphere

ionosphere

charged particles above the Earth

transmitter

reflection of radio waves

receiver

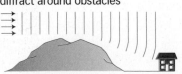

some radio waves are able to bend/diffract around obstacles

Microwaves

- Some microwaves pass **easily** through the Earth's atmosphere and so are used for communications via **satellites**, for example mobile phones.
- Microwaves are sometimes used for **cooking**, for example microwave ovens.
- Water molecules inside food absorb microwaves.
- They become 'hot' by vibrating vigorously, cooking the food from inside.

Microwaves can be dangerous if misused. They can cause **damage to living cells**.
Typical values
$\lambda = 3 \times 10^{-2}$ m, $f = 1 \times 10^{10}$ Hz.

satellite redirects (relays) signal

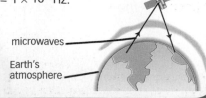

microwaves

food absorbs microwaves

Earth's atmosphere

Infra-red waves

- Infra-red waves are **given out by all warm objects**.
- Our skin is sensitive to infra-red waves.
- Over-exposure causes **sunburn** but not tanning.
- Infra-red waves are used to **see in the dark**. Special 'heat-seeking' cameras **create images** of objects using the **infra-red waves the objects are emitting**. These are often used by the emergency services to detect people trapped in collapsed buildings or lost in mountains or on moors.
- **Remote controls** for TVs, radios, etc. use infra-red waves to carry instructions.
  Typical values
  $\lambda = 1 \times 10^{-5}$ m, $f = 3 \times 10^{13}$ Hz.

Infra-red waves carry the instruction from the remote to the TV.

Visible light

- Our eyes make use of waves from this part of the spectrum to allow us to **see**.
- It is the one part of the electromagnetic spectrum to which our **eyes are sensitive**.
- Visible light is also used to send messages down **optical fibres**.
  Typical values
  $\lambda = 5 \times 10^{-7}$m, $f = 6 \times 10^{14}$Hz.

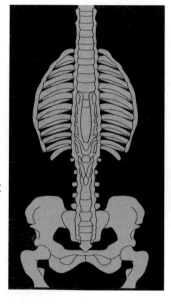

We use visible light to see.

Ultraviolet

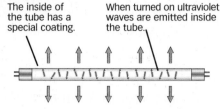

The inside of the tube has a special coating.

When turned on ultraviolet waves are emitted inside the tube.

When the ultraviolet waves strike this coating they are absorbed and visible light is emitted.

- Ultraviolet waves are **emitted** by the Sun.
- They cause our skin to tan.
- Over-exposure **can lead to skin cancer**.
- Certain chemicals **fluoresce (glow)** when exposed to ultraviolet.
- Words written with security markers are only visible in ultraviolet light.
  Typical values
  $\lambda = 3 \times 10^{-8}$m, $f = 1 \times 10^{16}$Hz.

X-rays

- X-rays have a **very short wavelength** and a **very high frequency**.
- They are **highly penetrating**.
- They are used to look for damaged bones inside the body.
- Over-exposure can cause cancer. Radiographers therefore stand behind **lead screens** or wear **lead aprons** to **prevent over-exposure** because X-rays cannot penetrate lead.
  Typical values
  $\lambda = 5 \times 10^{-10}$m, $f = 6 \times 10^{18}$Hz.

Gamma rays

- These are **very penetrating** waves that are emitted by some **radioactive materials** (see page 82).
- They can be used to kill harmful bacteria, for example to sterilise surgical equipment.
- If used correctly, they can be used to kill certain kinds of cancer. This procedure is called **radiotherapy**.
- Incorrect exposure or dosage can damage living cells and **cause cancer**.
  Typical values
  $\lambda = 5 \times 10^{-12}$m, $f = 6 \times 10^{21}$Hz.

Quick test

1 What properties are different for the different groups of waves?

2 What do all the highly penetrating waves have in common?

3 Name two properties that all these waves have in common.

4 Name three types of wave that can be used for communications.

5 Name three types of wave that might cause cancer.

6 Name one type of wave that can be used to treat cancer.

7 Name one source of gamma waves.

1. They have different wavelengths and frequencies. 2. They have high frequencies and short wavelengths. 3. They travel through vacuum at same speed, transverse, transfer energy (any two of) 4. visible, radio, microwave 5. ultraviolet, gamma, X-ray 6. gamma 7. radioactive materials

Exam-style questions   Use the questions to test your progress. Check your answers on pages 92–95.

1 a Starting with the longest wavelengths, write down in order the names of the main groups of waves which make up the electromagnetic spectrum.   [2]

....
....

b Name four properties that all these waves have in common.   [2]

....
....

c Describe a simple experiment you could carry out to confirm that waves in the electromagnetic spectrum are transverse waves.   [4]

....
....
....

2 a Explain in detail how emission line spectra can be used to identify the presence of a particular element in a gas.   [4]

....
....
....
....

b Explain in detail how absorption spectra can be used to identify the elements in the gas surrounding a star.   [4]

....
....
....

3 a What is a photon?   [2]

....
....

b Compare the energy of a photon of microwave radiation, wavelength 6.0×10^{-2} m, with that of a photon of X-ray radiation, wavelength 5.0×10^{-10} m.
($c = 3.0 \times 10^8$ m s^{-1}, Planck constant $h = 6.6 \times 10^{-34}$ J s)   [5]

....
....
....
....

4 a Explain briefly how you would use a gold-leaf electroscope to demonstrate photoelectric emission.   [3]

....
....

b Explain the meaning of the phrase 'zinc has a work function of 3.6 eV'.   [2]

....
....

c Calculate the threshold frequency for a metal that has a work function of 2.7 eV.
(Planck constant $h = 6.6 \times 10^{-34}$ J s)   [2]

....
....

5 a Sketch a graph to show the relationship between the maximum energy of photoelectrons and the frequency of the radiation which caused their emission. [3]

b A metal has a threshold frequency of 4.0×10^{14} Hz. Calculate the maximum energy of the photoelectrons emitted when light of frequency 6.0×10^{14} Hz is incident on the surface of the metal. [4]

....

....

....

c Explain why photoelectrons have a range of energies. [2]

....

....

6 a Explain the phrase 'wave–particle duality' when used to describe light. [2]

....

....

b Calculate the de Broglie wavelength of an electron moving with a speed of 1.8×10^7 m s^{-1}. (Planck constant $h = 6.6 \times 10^{-34}$ J s and mass of an electron $m_e = 9.1 \times 10^{-31}$ kg) [2]

....

....

7 The diagram shows the energy levels in a hydrogen atom.
a Explain what is meant by these phrases.
   i an electron is in the ground state [1]

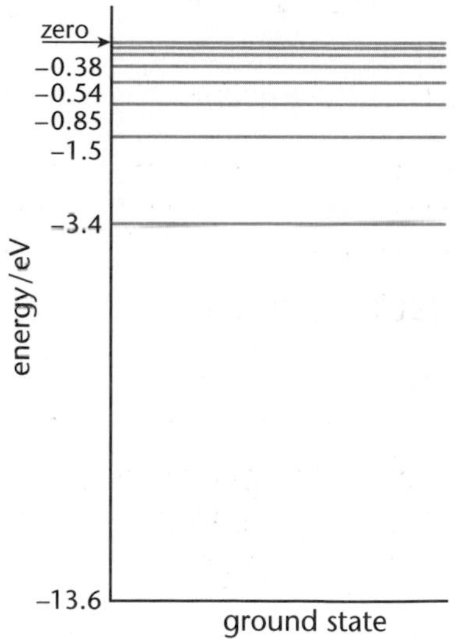

   ....

   ....

   ii an electron is in an excited state [1]

   ....

   ....

b How much energy does an electron need to receive in order to move from an energy level of −3.4 eV to an energy level of −1.5 eV? [1]

....

....

c How much energy will this electron lose if it falls to the ground state when at −1.5eV state?

....

d Calculate the wavelength of the light emitted by the electron in part c). ($c = 3.0 \times 10^8$ m s^{-1}, Planck constant $h = 6.6 \times 10^{-34}$ J s) [3]

....

....

....

Total   /50

energy/eV

zero
−0.38
−0.54
−0.85
−1.5

−3.4

−13.6
ground state

Energy levels in a hydrogen atom

Heat capacities

Specific heat capacity

If energy is transferred to an object such as a block of iron or a beaker of water, the increase in its internal energy is accompanied by an increase in temperature (unless it melts or boils).

Assuming there is no change in state, the relationship between the amount of energy given to the object (ΔQ) and the temperature rise ($\Delta\theta$) is described by the equation

$$\Delta Q = m \times c \times \Delta\theta$$

where m is the mass of the object and c is a constant for that material known as its <u>specific heat capacity</u>.

> The specific heat capacity of a substance is the amount of energy that must be given to 1 kg of the substance in order to increase its temperature by 1 K. It is measured in J kg^{-1} K^{-1}.

Here are some typical values for specific heat capacities

 or 1 °c

water

copper

aluminium

lead

Substance	Specific heat capacity (J kg^{-1} K^{-1})
Water	4200
Concrete	3400
Aluminium	910
Copper	390
Lead	130

EXAMPLE

Calculate the amount of energy that must be given to a 5.0 kg block of aluminium to increase its temperature from room temperature 15 °C to 100 °C.

$$\Delta Q = m \times c \times \Delta\theta$$
$$\Delta Q = 5 \times 910 \times 85$$
$$\Delta Q = 386\ 750 \text{ J or } 390 \text{ kJ}$$

Measurement of specific heat capacities

- The solid or liquid is heated electrically. The p.d. across the heater, the current flowing and the time the heater is switched on are all noted.
- The mass of the substance and the temperature rise caused by the heating are also noted.
- The energy supplied by the heater must be equal to the energy gained by the substance (assuming there is no energy loss to the surroundings).

$$IVt = m \times c \times \Delta\theta$$

therefore
$$c = \frac{IVt}{m \times \Delta\theta}$$

EXAMPLE

A 3.9 kg piece of copper was heated electrically until its temperature had risen by 5.0 °C. The p.d. across the heater was 10.0 V and the current passing through it was 2.0 A. If the heater was turned on for 400 s, calculate the specific heat capacity of copper.

$$c = \frac{IVt}{m \times \Delta\theta}$$
$$c = \frac{2.0 \times 10.0 \times 400}{3.9 \times 5}$$

$$c = 390 \text{ J kg}^{-1} \text{ K}^{-1}$$

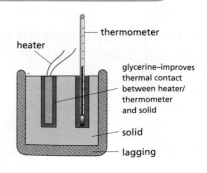

thermometer
heater
glycerine–improves thermal contact between heater/ thermometer and solid
solid
lagging

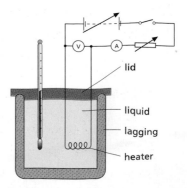

lid
liquid
lagging
heater

Heat capacities

Sometimes when objects are made of several different materials, it is sensible to consider the object as a whole rather than trying to take into account the proportions of the different materials. The **heat capacity** (C) of an object is the energy needed to increase the temperature of the object by 1 K.

$$\Delta Q = C\Delta\theta$$

Specific latent heat

- If an object melts, boils, condenses or freezes, we say it changes state. In order to change state it must gain or lose energy. This energy is called the **latent heat**.
- The **specific latent heat of fusion** (L_f) of a substance is the amount of energy necessary to change the state of 1 kg of that substance in the solid state into a liquid without any change in temperature.
- The **specific latent heat of vaporisation** (L_v) of a substance is the amount of energy necessary to change the state of 1 kg of that substance in the liquid state into a gas without any change in temperature.

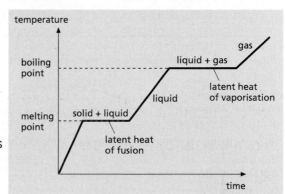

The relationship between the heat supplied (ΔQ) and the mass of the substance, m, which changes state is described by the equation

$$\Delta Q = m \times L$$

L is measured in J kg^{-1}.
Here are some typical values for specific latent heat capacities.

Substance	L_f / kJ kg^{-1}	L_v /kJ kg^{-1}
Water	330	2 260
Aluminium	380	10 800
Copper	210	4 700
Mercury	16	290

EXAMPLE
Calculate the energy necessary to melt 40 kg of aluminium with no change in temperature.
$\Delta Q = m \times L = 40 \times 380$
$= 15\ 200$ J or 15 kJ

Quick test

1 Explain the statement 'The specific heat capacity of copper is 390 J kg^{-1}K^{-1}'.

2 Explain the statement 'The heat capacity of a kettle is 1000 J K^{-1}'.

3 Calculate the energy that must be given to a 5.0 kg block of lead in order to increase its temperature from 15 °C to 65 °C.

4 Calculate the energy needed to change 2.5 kg of water at 100 °C into steam at 100 °C.

5 A 300 W heater is turned on for 40 s. The energy it releases causes a 1.5 kg block of a solid to increase its temperature from 15 °C to 35 °C. Assuming no heat is lost to the surroundings, calculate the specific heat capacity of the solid.

1. 390 J of energy are needed to increase the temperature of 1 kg of copper by 1 K. 2. 1000 J of energy are needed to increase the temperature of the kettle by 1 K. 3. 32 500 J or 33 kJ 4. 5.7 MJ 5. 400 J kg^{-1} K^{-1}

67

The ideal gas laws

Pressure in a gas

The gas atoms/molecules inside this container are <u>continually bouncing off the sides</u>. It is these collisions with the sides of the container that <u>create the pressure</u> inside the container. There are several factors that will affect the size of the pressure:

- **the mass of the gas in the container**
- **the volume of the container**
- **the temperature of the gas**

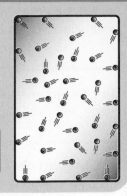

Mass

If **more gas atoms/molecules** are pushed into the container, i.e. the mass of the gas is increased, there will be **more frequent collisions** and so the **pressure of the gas will increase**.

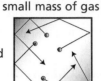

small mass of gas — Few collisions per second therefore low pressure.

larger mass of gas — More collisions per second therefore higher pressure.

Volume

If the **volume** of the container is **decreased** the atoms/molecules will have less distance to travel between collisions. There will therefore be **more collisions per second**, i.e. the **pressure of the gas will increase**. If the number of gas atoms/molecules in a container and its temperature are kept constant the relationship between the pressure of the gas and its volume is described by the equation

$$P_1V_1 = P_2V_2$$

This is known as **Boyle's law**.

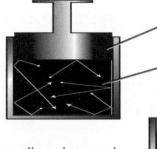

low pressure

large volume

smaller volume and more collisions per second therefore higher pressure

Temperature

low temperature   higher temperature

few collisions per second therefore low pressure

more collisions per second therefore higher pressure

If the **temperature** of a gas is **increased** its atoms/molecules will move faster, there will be **more collisions** each second and there will again be an **increase in pressure**. If the number of atoms/molecules in the container and the volume of the container are kept constant, the relationship between the pressure and the temperature of the gas is described by the equation

$$\frac{P_1}{T_1} = \frac{P_2}{T_2}$$

This is known as the **Pressure law**.
For this equation to work, the temperatures must be measured using the **Kelvin scale**. To convert degrees Celsius to Kelvin, we add **273 K**.

EXAMPLE

A sample of gas at a pressure of 200 kPa and at a temperature of 27 °C is heated to 327 °C. Calculate the new pressure of the gas assuming the mass and volume of the gas did not change.

$\frac{P_1}{T_1} = \frac{P_2}{T_2}$ $T_1 = 273 + 27 = 300$ K
            $T_2 = 273 + 327 = 600$ K

$\frac{200}{300} = \frac{P_2}{600}$

$P_2 = 2 \times 600/300 = 400$ kPa

Charles' Law

If a sample of gas is **heated** at **constant pressure**, the **number of collisions per second increases** and the gas **expands**. If the number of particles in the sample and its pressure are kept constant, the relationship between the volume of the gas and its temperature is described by the equation

$$\frac{V_1}{T_1} = \frac{V_2}{T_2}$$

This is known as **Charles' law**.

Again, for this equation to work, the temperatures must be measured in Kelvin.

EXAMPLE

100 cm³ of a gas at a temperature of 27 °C is heated to a temperature of 627 °C at constant pressure. Calculate the volume of the gas at this temperature.

$$\frac{V_1}{T_1} = \frac{V_2}{T_2}$$

$T_1 = 273 + 27 = 300$ K
$T_2 = 273 + 627 = 900$ K

$$\frac{100}{300} = \frac{V_2}{900}$$

$$V_2 = 300 \text{ cm}^3$$

Ideal gas equation

These three gas laws can be combined together into one equation.

$$\frac{P_1 V_1}{T_1} = \frac{P_2 V_2}{T_2} \quad \text{or} \quad \frac{PV}{T} = \text{constant}$$

The value of the constant depends upon the amount of gas present.

But for 1 mole* of gas it has the value 8.3 J mol⁻¹ K⁻¹.
It is known as the molar gas constant (R). This equation can therefore be written as

$$PV = nRT$$

where n is the number of moles of gas present

*1 mole of gas contains 6.02×10^{23} atoms/molecules.

EXAMPLE

Calculate the volume of 3.0 moles of an ideal gas that is at a temperature of 27 °C and at a pressure of 1.0×10^3 Pa.

$$PV = nRT$$

$$V = \frac{nRT}{P}$$

$$V = \frac{3.0 \times 8.3 \times 300}{1.0 \times 10^3}$$

$$V = 7.5 \text{ m}^3$$

Quick test

1 Explain why letting some air out of a tyre decreases the pressure.

2 Explain why a sealed container of gas may explode if heated.

3 The pressure of a fixed mass of gas at a constant temperature is trebled. What effect does this have on the volume of the gas?

4 Calculate the final pressure of a sample of gas which initially has a volume of 100 cm³ at a pressure of 100 kPa and a temperature of 27 °C but is then compressed to a volume of 50 cm³ and heated to a temperature of 327 °C.

5 Calculate the number of moles there are in a sample of gas that has a volume of 3.0 m³ at a pressure of 2.0×10^3 Pa and a temperature of 600 K.

1. fewer molecules, therefore fewer collisions, so does the rate of the collisions with the sides of the container, i.e. the pressure of the gas increases 2. As the temperature increases 3. The volume of the gas will now be 1/3 the original volume. 4. 400 kPa 5. 1.2 mol

69

Mathematical description of gases

- The kinetic theory helps us to explain some of the properties of gases such as diffusion, pressure, etc. We can extend these ideas to produce an equation that will describe some of these properties quantitatively.

Before we can derive the equation, we need to make certain assumptions about the way atoms/molecules of a gas behave.

- The atoms/molecules of a gas are in continuous and random motion. This assumption is supported by Brownian motion.

- The total volume of the gas atoms/molecules is negligible compared with the volume of the gas itself. This is true for all gases at low pressures.

An ideal gas

- The atoms/molecules of a gas undergo perfectly elastic collisions, i.e. there is no loss in kinetic energy.

- There are no intermolecular forces except during collisions between atoms and the container walls.

A gas for which all the above assumptions are true is described as being an ideal gas.

If all the above assumptions hold, we can show that the pressure of an ideal gas is directly proportional to the mean of the squares of the speeds of the atoms/molecules.

$$PV = \tfrac{1}{3} Nm\ \overline{c^2}$$

where N is the number of atoms/molecules in the gas, m is the mass of each particle and $\overline{c^2}$ is the mean of the squares of the speeds of the atoms/molecules.

Nm is the total mass of the gas and V is the volume it occupies.

But density (ρ) = mass / volume

Therefore we can write the above equation in the form

$$P = \tfrac{1}{3} \rho\ \overline{c^2}$$

EXAMPLE

Calculate the pressure of a gas whose atoms have a root mean square speed of 500 m s⁻¹ and a density of 1.2 kg m⁻³.

Using $P = \tfrac{1}{3} \rho\ \overline{c^2}$

$P = \tfrac{1}{3} \times 1.2 \times 500^2$

$P = 1.0 \times 10^5$ Pa

EXAMPLE

Calculate the root mean square speed for five atoms in a gas which have speeds 2 m s⁻¹, 3 m s⁻¹, 4 m s⁻¹, 5 m s⁻¹ and 6 m s⁻¹.

$$\overline{c^2} = \frac{4 + 9 + 16 + 25 + 36}{5} = \frac{90}{5}$$

$$\sqrt{\overline{c^2}} = 4.2 \text{ m s}^{-1}$$

Relationship between kinetic energy and temperature

Consider one mole of gas. This contains 6.02×10^{23} atoms or molecules. This is the **Avogadro constant** (N_A).

We can write $\quad PV = \frac{1}{3} N_A m \overline{c^2}$

where m is the mass of one molecule of the gas, but it is also true for one mole of gas that

$PV = RT$

therefore $\qquad \frac{1}{3} N_A m \overline{c^2} = RT$

or $\qquad \frac{1}{3} m \overline{c^2} = \dfrac{RT}{N_A}$

or $\qquad \frac{1}{3} m \overline{c^2} = kT$

where k is a constant known as the **Boltzmann constant**. The Boltzmann constant has a value of 1.38×10^{-23} J K^{-1}.

If we multiply both sides of this equation by 3 and divide by 2, we can rewrite this equation in a more useful form.

$$\frac{1}{2} m \overline{c^2} = \frac{3}{2} kT$$

We can see from this equation that **the mean kinetic energy of the gas atoms or molecules is directly proportional to the temperature of the gas expressed in Kelvin**.

Molecular speeds in gases

The molecules of a gas have a wide spread (or range) of velocities. The graph shows how we believe these speeds to be distributed statistically.

- The molecules can theoretically have speeds from zero to infinity.
- At any temperature, there are very few particles with very low speeds or very high speeds.
- If the temperature of a gas increases, more molecules will have higher speeds.

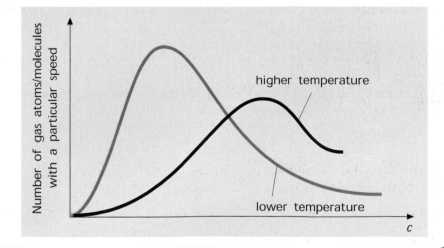

higher temperature

lower temperature

Number of gas atoms/molecules with a particular speed

c

Quick test

1. What is an ideal gas?
2. What would happen to the molecules of a gas if their collisions were not elastic?
3. Which two assumptions might not be true for a gas at high pressure?
4. Calculate the density of a sample of gas at a pressure of 1.3×10^5 Pa if the root mean square speed of its particles is 400 m s^{-1}.
5. Describe what happens to the mean kinetic energy of a sample of gas at a fixed volume whose temperature increases from 27 °C to 927 °C.

1. a gas that obeys the gas laws 2. they would lose energy and, eventually, stop. 3. The volume of the molecules is negligible compared with the volume of the gas and there are no intermolecular forces between the molecules other than when they collide. 4. 2.4 kg m^{-3} 5. It increases four-fold.

71

Heat and work

Temperature difference and the flow of heat

Heat naturally flows from an object at a higher temperature to an object at a lower temperature. If there is no overall flow of heat between two objects that are in thermal contact, the objects are at the same temperature and are in <u>thermal equilibrium</u>.

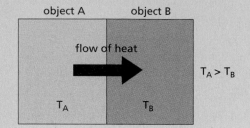

Internal energy

- The atoms in a solid, a liquid or a real gas have both potential and kinetic energy.
- The distribution of these energies within an object is continually changing.
- The sum of the kinetic and potential energies of the atoms within an object is known as its <u>**internal energy**</u>.

First law of thermodynamics

The internal energy of an object can be increased by supplying it with heat energy or by doing work on it.

This is summarised in the <u>**first law of thermodynamics**</u>, which states that the change in internal energy of a system (ΔU) is equal to the sum of the energy entering the system by heating (ΔQ) and the energy entering the system by work being done on it (ΔW).

$$\Delta U = \Delta Q + \Delta W$$

EXAMPLE

Calculate the increase in the internal energy of a gas that gains 50 J of energy from a heat source and then has 100 J of work done on it as it is compressed.

$\Delta U = \Delta Q + \Delta W$

$\Delta U = 50 + 100$

$\Delta U = 150$ J

If this piston is pushed down, work is done on the gas and its internal energy will increase. If no heat is allowed to escape, the temperature of the gas will increase, i.e. ΔQ is zero, ΔW is positive and so ΔU is positive.

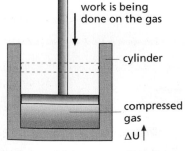

If the gas in this cylinder is at a higher pressure than atmospheric pressure, it may push the piston upwards, i.e. the gas is now doing work. The internal energy of the gas will therefore decrease. If no heat is allowed to enter the cylinder, the temperature of the gas will fall, i.e. ΔQ is zero, ΔW is negative and so ΔU is negative.

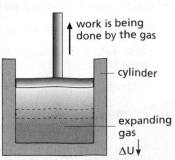

Heat engines

Petrol, diesel and jet engines are examples of **heat engines**. They convert heat into work.

- A fuel is burnt to create a source of high temperature.

- Heat flows from this source towards a cooler material (often the atmosphere).

- Some of this energy is used to do work, for example move pistons.

- Heat that is not used to do work flows into the cooler material, i.e. it is transferred to the surroundings or to a heat sink.

It is not possible to convert all the heat that flows from the high temperature source into work and therefore heat engines can never be 100% efficient.

The theoretical efficiency of a heat engine is given by this equation.

$$\frac{\text{work done by engine}}{\text{heat supplied to engine}} = \frac{W}{Q_1} = \frac{Q_1 - Q_2}{Q_1} = \frac{T_1 - T_2}{T_1}$$

where T_1 is the temperature of the heat source in K.

T_2 is the temperature of the heat sink in K.

EXAMPLE

Calculate the maximum efficiency of a heat engine that operates with a heat source at 327 °C (600 K) and a heat sink at 27 °C (300 K).

$$\text{Maximum efficiency} = \frac{T_1 - T_2}{T_1}$$
$$= \frac{600 - 300}{600}$$
$$= 0.5 \text{ or } 50\%$$

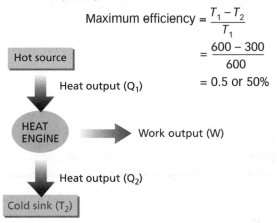

Heat pumps

Heat pumps move heat from places of low temperature to places of higher temperature. For example, a heat pump could be used to remove heat from a freezer and transfer it to the surroundings.

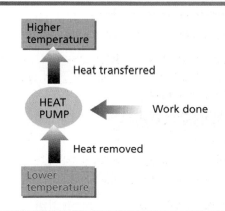

Quick test

1 In which direction will heat naturally flow?

2 What is the internal energy of an object?

3 Suggest two ways in which you could increase the internal energy of a gas.

4 Give one example of how work

   a could be done on a gas

   b could be done by a gas

5 What is a heat engine?

6 What is a heat pump?

7 Calculate the change in the internal energy of a gas when 200 J of heat are taken from it and 150 J of work are done on it.

8 Calculate the maximum efficiency of a heat engine which operates between a source at 627 °C and the surroundings at 27 °C.

1. from a hotter place to a cooler place 2. the sum of the kinetic and potential energies of its atoms 3. supply heat energy to it and do work on it 4. a) compressing the gas b) allowing the gas to expand 5. A device that converts heat into work 6. A device that moves heat from places of lower temperature to places of higher temperature 7. The gas loses 50 J of energy 8. 0.67 or 67%

Hooke's law and the Young modulus

If a tensile force is applied to a wire (or spring) it will stretch. How much the wire stretches depends on:

- its size and shape, i.e. its length and cross-sectional area

How much it will stretch

- the size of the applied force

- the material from which the wire is made.

The graph below shows how the extension of a wire changes as the applied force is increased.

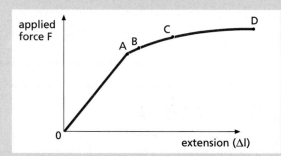

- Between O and A the extension of the wire (Δl) is directly proportional to the force (F) applied to it, i.e. $F = k\,\Delta l$ where k = force constant, or spring constant. In this region of the graph the wire is said to be obeying Hooke's law.

The extension of the wire is directly proportional to the forces that produce it.

- **A** is known as the **limit of proportionality**. Beyond this point, the extension of the wire is no longer proportional to the forces applied to it.
- **B** is known as the **elastic limit**. Beyond this point, the applied forces will cause permanent deformation of the wire, i.e. when the forces are removed the wire will not return to its original dimensions.
- **C** is known as the **yield point**. Beyond this point, small increases in the applied force will produce large increases in extension.
- At D, the wire breaks.

Stress and strain

Stress (σ) is the force per unit cross-sectional area applied to an object. It is measured in N m^{-2} or Pa.

$$\sigma = \frac{F}{A}$$

EXAMPLE

Calculate the stress experienced by a wire of cross-sectional area 2.0×10^{-6} m^2 when a force of 100 N is applied to it.

$$\sigma = \frac{F}{A} = \frac{100}{2.0 \times 10^{-6}} = 5.0 \times 10^7 \text{ Pa}$$

Strain (ε) is the change in length per unit length. Strain has no units. It is a ratio.

$$\varepsilon = \frac{\Delta l}{l}$$

EXAMPLE

Calculate the strain experienced by a wire 2.5 m long when a tensile stress causes it to stretch 1.25 mm.

$$\varepsilon = \frac{\Delta l}{l} = \frac{1.25 \times 10^{-3}}{2.5} = 5 \times 10^{-4}$$

The Young modulus

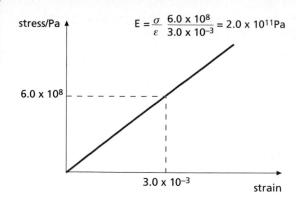

$E = \dfrac{\sigma}{\varepsilon}\ \dfrac{6.0 \times 10^8}{3.0 \times 10^{-3}} = 2.0 \times 10^{11} \text{Pa}$

- If a stress is applied to a wire it will produce a strain.
- Providing the limit of proportionality is not exceeded, the strain produced will be proportional to the applied stress.
- The ratio of the applied stress to the strain produced is equal to the gradient of the graph and is constant for the material.
- This constant is known as the **Young modulus** (E) and has the units N m^{-2} or Pa.
- The greater the Young modulus, the more rigid the material.

Here are some typical values of the Young modulus.

Material	Young modulus / Pa
Steel	2.1×10^{11}
Copper	1.1×10^{11}
Aluminium	7.0×10^{10}
Nylon	3.6×10^{9}

Calculating the Young modulus

We can calculate the Young modulus for a material using this equation:

$$E = \frac{\sigma}{\varepsilon}$$

Expanding this equation we can write

$E = \dfrac{F/A}{\Delta l / l}$ or $E = \dfrac{Fl}{A\Delta l}$

EXAMPLE

A steel wire 2.5 m long and with a cross-sectional area of 0.01 cm^2 stretches by 0.50 cm when a tensile force of 400 N is applied to it.
Calculate the Young modulus for steel.

$E = \dfrac{Fl}{A\Delta l} = \dfrac{400 \times 2.5}{0.01 \times 10^{-4} \times 0.50 \times 10^{-2}} = 2.0 \times 10^{11}$ Pa or 200 GPa

Energy stored in a stretched wire

- When a wire is being stretched, work is being done and energy is stored in the wire.
- If the elastic limit has not been exceeded and no energy is lost to the surroundings, the work done on the wire will be equal to the elastic potential energy gained by the wire.
- **The energy stored in the wire** = $\frac{1}{2} F \Delta l$ where F is the final force applied to the wire, and Δl = final extension.
- Substituting for F from the Young modulus equation, we can derive a second equation.

energy stored per unit volume of wire $= \frac{1}{2} \times$ **stress** $\times$ **strain**

EXAMPLE

Calculate the energy stored per unit volume when a stress of 2.0×10^{10} Pa produces a strain of 5.0×10^{-2}.

energy stored $= \frac{1}{2} \times$ stress $\times$ strain
$= \frac{1}{2} \times 2.0 \times 10^{10} \times 5.0 \times 10^{-2}$
$= 5.0 \times 10^{8}$ Jm^{-3}

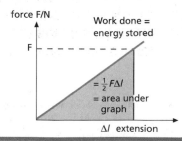

Quick test

1 What is a tensile force?

2 What is meant by the term 'elastic limit'?

3 Calculate the stress created when a force of 100 kN is applied to a wire of cross-sectional area 0.04 cm^2.

4 Calculate the strain experienced by a 2.0 m length of steel wire that extends by 0.04 mm when a tensile force is applied to it.

5 Calculate the Young modulus for a material which experiences a strain of 2.5×10^{-1} when a stress of 5.0×10^{9} Pa is applied to it. You may assume that the limit of proportionality has not been exceeded.

1. a stretching force 2. The elastic limit is the maximum force that can be applied to an object without causing permanent deformation. 3. 2.5×10^{10} Pa 4. 2.0×10^{-4} 5. 2.0×10^{10} Pa

75

Materials

● **Stress–strain graphs can be very useful tools in helping engineers and designers to analyse important physical properties of different materials.**

Ductile materials

- Materials such as copper, steel and aluminium are ductile.
- Their stress–strain graphs show that they yield before they break.
- They can therefore be shaped or permanently deformed without cracking.
- Elastic deformation takes place before the yield point.
- Plastic deformation takes place after the yield point.
- Because they are ductile, materials like these can be drawn into wires.

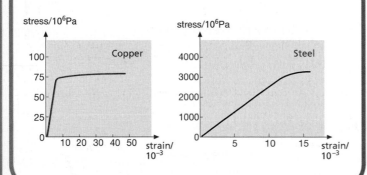

Brittle materials

- Materials such as glass, cast iron and concrete are brittle.
- There is no observable yield point.
- They undergo elastic deformation, obeying Hooke's law.
- They break without showing any plastic deformation.
- Because they are brittle materials, they cannot easily be shaped without cracking.

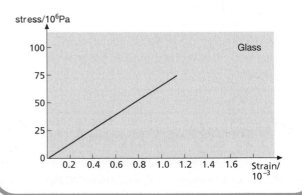

Explaining the behaviour of materials

Explaining the behaviour of materials when stresses are applied to them

The graph shows the forces between neighbouring particles in a solid.

- There are both attractive and repulsive forces between neighbouring particles in a solid.
- At the equilibrium position these forces are balanced.
- If a particle is moved from its equilibrium position, the forces acting upon it will change.
- If a compressive force tries to push the particles closer together, the repulsive forces increase.
- If a tensile force tries to pull the particles further apart, the attractive forces increase.
- For small tensile forces, the increase in the attractive forces is proportional to the increase in separation, i.e. the material is obeying Hooke's law.
- If these small tensile forces are removed, the particles will return to their original positions. There is **no permanent deformation**. The material has undergone **elastic deformation**.

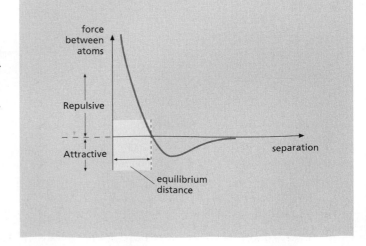

- If larger tensile forces are applied, the attractive forces between neighbouring particles may be overcome, resulting in a rearrangement of the particles.
- If the tensile forces are now removed, the particles do not return to their original positions. There is **permanent deformation**. The material has undergone **plastic deformation**.

Stretching rubber

- The shape of a stress–strain graph for rubber is quite different from the materials we have looked at so far.
- Its shape indicates that at first it is hard to stretch, then it becomes easier, and then it becomes harder again.
- Before stressing, rubber consists of long-chained molecules which are all tangled, making it difficult to stretch.
- Once the initial stretching has untangled the long-chained molecules, further stretching is much easier as the molecules become aligned with the applied stress.
- Having become aligned, further stretching now requires the bonds within the molecules to be stretched. This requires higher stresses.

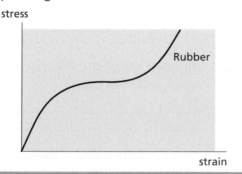

The stress–strain graph below shows what happens to a ductile material if it is stretched beyond its elastic limit.

Work is done in stretching the material. But when the stress is removed, only part of the energy stored is recovered. The remainder of the energy is lost.

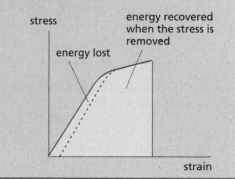

Hysteresis loop for rubber

- When rubber is stretched it absorbs energy.
- When the applied force is removed, the rubber releases energy.
- Rubber absorbs more energy as it is stretched than it releases as it is unloaded.
- This behaviour can be seen in the stress–strain graph for rubber.
- This loop is called a **hysteresis loop**.
- The area within the loop represents the energy lost (not recovered) per unit volume of the rubber for each cycle of loading and unloading.
- This lost energy is likely to lead to an increase in the temperature of the rubber.

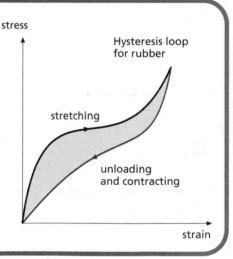

Quick test

1 What is a ductile material?

2 What is a brittle material?

3 Sketch a stress–strain graph for

    a a ductile material

    b a brittle material.

4 Explain the difference between elastic deformation and plastic deformation.

5 Explain briefly why there are three different sections in the stress–strain curve for rubber.

1 a Explain what is meant by the statement 'water has a specific heat capacity of 4200 J kg⁻¹ K⁻¹'. [1]

....

b Describe briefly how you would determine a value for the specific heat capacity of a liquid. [4]

....

....

....

c Calculate the energy needed to change 2.0 kg of ice at 0 °C to steam at 100 °C.
(L_f ice 300 kJ kg⁻¹, L_v water 2260 kJ kg⁻¹) [6]

....

....

....

....

2 a Explain why increasing the temperature of a fixed mass of gas at constant volume results in an increase in pressure. [2]

....

....

b Sketch the pressure–temperature graph for a fixed mass of gas at constant volume. [2]

c A gas of volume 2.0×10^{-4} m³ is at a temperature of 127 °C is heated to a temperature of 527 °C at constant pressure. Calculate the volume of the gas at 527 °C. [2]

....

....

d If the pressure of the gas in part c) is 1.0×10^5 Pa, calculate the number of moles present.
($R = 8.3$ J mol⁻¹ K⁻¹) [3]

....

....

....

3 For an ideal gas $PV = \frac{1}{3} Nm\overline{c^2}$
a Explain the meaning of each of the symbols in this equation. [5]

....

....

....

....

b State the assumptions made in the derivation of the above equation. [4]

....

....

....

c Calculate the root mean square speed for four particles in a gas which have speeds of 300 m s⁻¹, 350 m s⁻¹, 400 m s⁻¹ and 450 m s⁻¹. [2]

....

....

d Calculate the pressure exerted by a gas whose atoms have a root mean square speed of 550 m s^{-1} and a density of 1.5 kg m^{-3}. [2]

....

....

4 The diagram below shows apparatus that can be used to measure the Young modulus for a wire.

a What measurements should be taken to determine the stress applied to the wire? [1]

support

reference wire

test wire

vernier gauge

small weight to keep wire taut

load

....

....

b What measurements should be taken to determine the strain experienced by the wire? [1]

....

....

c If the stress applied to the wire is 3.0 × 10^8 Pa and this causes a strain of 2.0 × 10^{-3}, calculate the Young modulus for the wire. [2]

....

....

d calculate the energy stored per unit volume of the wire. [2]

....

....

5 Look carefully at the force–extension graph drawn below.

a Describe the behaviour of the ductile material whose force-extension graph is shown here. [2]

P = limit of proportionality
E = elastic limit
B = breaking point

....

....

....

....

b Give one example of a ductile material. [1]

....

6 The first law of thermodynamics can be described by the equation
$$\Delta U = \Delta Q + \Delta W$$

a Explain the meaning of each of the symbols in the above equation. [3]

....

....

....

b Explain the difference between a heat engine and a heat pump. [2]

....

....

c Calculate the maximum efficiency of a heat engine that takes heat from a source at a temperature of 227 °C and delivers it to a heat sink at a temperature of 27 °C. [2]

....

....

Total   /49

Atomic structure

For many years scientists thought that atoms were the smallest particles that could exist. With a greater understanding of electricity it was suggested that atoms must contain positive and negative charges. A scientist called J J Thompson suggested that atoms have a positive body, which contains negative particles, like a pudding that contains plums. But an experiment carried out by scientists called Rutherford, Geiger and Marsden suggested a different model. This new model was rapidly accepted as being correct. The experiment they carried out was called the **alpha scattering experiment**.

The alpha scattering experiment

- Small particles called **alpha particles** were aimed at a **very thin piece of gold foil**.

- Most of the alpha particles **passed straight through the foil** and were not deviated.

- Some particles were deviated a little.

- A very small number of alpha particles were **scattered backwards**.

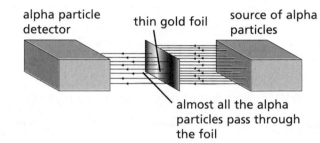

alpha particle detector   thin gold foil   source of alpha particles

almost all the alpha particles pass through the foil

From these results it was concluded that:
- most of an atom is **empty space**
- most of the mass of an atom is concentrated in a **very small central nucleus** which must be positively charged.

The model became known as the **nuclear atom**.

The nuclear atom

Particle	Location	Relative atomic mass	Relative charge
Proton	In nucleus	1	+1
Neutron	In nucleus	1	0
Electron	Around nucleus	$(\frac{1}{2000})$	−1

- Atoms have **no overall charge**. They are **neutral**.
- They must therefore **contain equal numbers of protons and electrons**.
- The number of protons an atom has in its nucleus is called the **proton number** or the **atomic number** (Z).
- The number of protons and neutrons an atom has in its nucleus is called the **nucleon number** or **mass number** (A).
- **The nucleus of an atom can be represented as $^A_Z X$, where x is the chemical symbol for the element.**

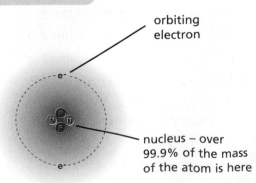

orbiting electron

nucleus – over 99.9% of the mass of the atom is here

Drawing atomic nuclei

Helium
- Helium has a proton number of 2. Its nucleus contains two protons.
- Helium has a nucleon number of 4. Its nucleus contains two protons and (4 – 2) neutrons.

Lithium
- Lithium has a proton number of 3. Its nucleus contains three protons.
- Lithium has a nucleon number of 7. Its nucleus contains three protons and (7 – 3) neutrons.

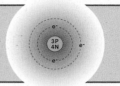

Potassium
- Potassium has a proton number of 19. Its nucleus contains 19 protons.
- Potassium has a nucleon number of 39. Its nucleus contains 19 protons and (39 – 19) neutrons.

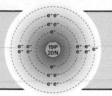

Isotopes

- Some atoms of the same element have **nuclei** with **different numbers of neutrons**.
- For example, all atoms of chlorine have 17 protons in their nuclei, but some atoms have 18 neutrons in their nuclei whilst others have 20.
- These different forms of the same element are called **isotopes**.
- Isotopes are nuclei that contain the same number of protons, but a different number of neutrons

The two isotopes of chlorine can be written as chlorine–35 and chlorine–37.

A nucleus with a particular structure of protons and neutrons is referred to as a **nuclide**.

this isotope is called chlorine –37

this isotope is called chlorine –35

Quick test

1 What observations from the alpha scattering experiment suggested these?

  a most of an atom is empty space

  b the nucleus of an atom is positively charged.

2 Write down the relative atomic mass of these.

  a a proton   **b** an electron   **c** a neutron

3 What is the proton number of a nucleus?

4 What is the nucleon number of a nucleus?

5 What are isotopes?

6 Compare the number of protons in the nucleus of an atom with the number of electrons orbiting it.

7 Describe the nucleus of these atoms.

  a $_8^{16}O$ **b** $_{13}^{27}Al$ **c** $_{30}^{64}Zn$

1.a) Most of the alpha particles were undeviated.   b) Back scattering was caused by electrostatic repulsion between the positively charged alpha particles and the nucleus.   2.a) 1   b) $\approx \frac{1}{2000}$ c) 1   3. the number of protons in the nucleus.   4. the number of protons + the number of neutrons in the nucleus   5. Isotopes are atoms of the same element that have different numbers of neutrons in their nuclei.   6. The number of protons is equal to the number of electrons.   7.a) The oxygen nucleus contains 8 protons and 8 neutrons.   b) The aluminium nucleus contains 13 protons and 14 neutrons.   c) The zinc nucleus contains 30 protons and 34 neutrons.

Radioactivity

- The nuclei of some atoms are unstable. In order to become more stable they emit either a particle or/and electromagnetic radiation. We describe such nuclei as being <u>radioactive</u>.
- There are radioactive materials all around us. Some of them are artificial and used in hospitals, nuclear power stations and even in the home. But most radioactive materials occur naturally. They are in the ground, in the food we eat, they are even in the air we breathe. Some radiation reaches us from space.
- The sum of the radiation produced by all these sources is called <u>background radiation</u>.
- There are <u>four</u> main types of radiation that may be emitted by a nucleus.

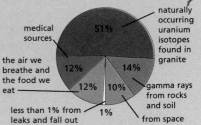

medical sources

naturally occurring uranium isotopes found in granite — 51%

the air we breathe and the food we eat — 12%

14% gamma rays from rocks and soil

12% 10%

less than 1% from leaks and fall out — 1%

from space

Alpha radiation (α)

- These are <u>slow-moving helium nuclei</u>, i.e. they consist of <u>two protons and two neutrons</u>.
- They have <u>poor penetration</u> (just a few centimetres in air).
- They collide with lots of atoms, <u>knocking some of their electrons off</u> and <u>creating ions</u>. They are <u>very good ionisers.</u>
- They are <u>positively charged</u> and so can be <u>deflected by electric and magnetic fields</u>.

This equation describes the emission of an alpha particle from a uranium–238 nucleus:

$$^{238}_{92}\text{U} \rightarrow \, ^{234}_{90}\text{Th} + \, ^{4}_{2}\text{He}$$

The atom of uranium–238 is <u>transmuted</u> into an atom of thorium–234 by alpha emission. The nucleus of the thorium–234 is described as the <u>daughter nuclide</u>.

Note In nuclear reactions the proton number and the nucleon number are conserved.

Beta-minus radiation (β–)

- These are <u>fast-moving, negatively charged electrons</u>.
- They have <u>quite good penetrating powers</u> (up to about a metre in air).
- They do collide with atoms and produce ions but not as many as the alpha particles.
- Because they are <u>negatively charged</u> they can be <u>deflected by electric and magnetic fields</u>.

This equation describes the emission of a beta-minus particle from a carbon–14 nucleus:

$$^{14}_{6}\text{C} \rightarrow \, ^{14}_{7}\text{N} + \, ^{0}_{-1}\text{e} + \nu_e$$

The beta particle is created in the nucleus when a neutron changes into a proton, an electron and an antineutrino. The proton remains in the nucleus and the electron is emitted as a fast-moving beta particle.

$$^{1}_{0}\text{n} \rightarrow \, ^{1}_{1}\text{p} + \, ^{0}_{-1}\text{e} + \bar{\nu}_e$$

Beta-plus radiation (β+)

These are fast-moving, positively charged electrons, more properly known as <u>positrons</u>.

This equation describes the emission of a beta-plus particle from a sodium–22 nucleus.

$$^{22}_{11}\text{Na} \rightarrow \, ^{22}_{10}\text{Ne} + \, ^{0}_{+1}\text{e} + \nu_e$$

Beta-plus emission occurs when a proton in the nucleus changes into a neutron, a positron and a neutrino.

$$^{1}_{1}\text{p} \rightarrow \, ^{1}_{0}\text{n} + \, ^{0}_{+1}\text{e} + \nu_e$$

The positron is an <u>antimatter particle</u>. If it meets an electron, they are both annihilated, leaving pure electromagnetic energy.

Gamma rays (γ)

After emitting an alpha or beta particle, the nucleus of an atom may have excess or residual energy. This is removed by the emission of a photon of electromagnetic radiation called gamma radiation. The photons have no mass or charge and so do not affect the proton number or the nucleon number of the nucleus.

- Gamma rays are <u>short-wavelength electromagnetic waves</u>, similar to X-rays.
- They <u>travel at the speed of light</u> and are <u>very penetrating</u>. (They can travel almost unlimited distances through air.)
- They are <u>very poor ionisers</u>.
- Gamma radiation <u>carries no charge</u> and so is <u>unaffected by magnetic and electric fields</u>.

Comparing the properties of radiation

Alpha, beta and gamma radiation

Radiation	α	β	γ
Mass	4	$(\frac{1}{2000})$	0
Relative charge	+2	−1	0
Relative ionising power	100 000	1000	1
Approximate penetrating power in air	1–5 cm	10–80 cm	Almost unlimited

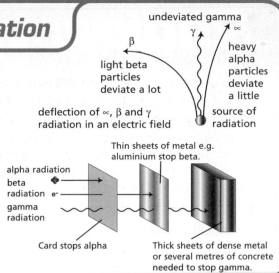

undeviated gamma

β
light beta particles deviate a lot

γ

∝
heavy alpha particles deviate a little

deflection of ∝, β and γ radiation in an electric field

source of radiation

Thin sheets of metal e.g. aluminium stop beta.

alpha radiation
beta radiation e-
gamma radiation

Card stops alpha

Thick sheets of dense metal or several metres of concrete needed to stop gamma.

Exposure to radiation

- **Absorption** of any of the three types of radiation by living cells is potentially dangerous.
- Absorption may cause **cell damage** and lead to illnesses such as **cancer**.
- Higher levels of exposure to these radiations may **kill living cells**.
- Those at most risk, for example radiographers, wear **radiation badges**. These contain photographic film which, when developed, shows the **degree of exposure** to radiation for that worker.

Exposure from sources outside the body
- Alpha radiation is the least dangerous as it is the least penetrating and unlikely to pass through clothing or skin.
- Beta and gamma radiations are able to pass through clothing and skin and, therefore, are potentially more dangerous.

Exposure from sources inside the body
- Alpha is the most dangerous radiation as it is most strongly absorbed by living cells and therefore causes most damage.
- Beta and gamma are, comparatively, not so dangerous as they are less likely to be absorbed by living cells.

Quick test

1 What is background radiation?

2 Why do some nuclei emit radiation?

3 Which of the four different types of radiation from radioactive materials

   a is positively charged

   b is unaffected by magnetic fields

   c is a very good ioniser

   d has a relative atomic mass of almost zero

   e consists of antimatter particles?

4 Write a nuclear equation which describes a nucleus decaying by emitting

   a an alpha particle

   b a beta-minus particle.

5 Explain why alpha radiation is potentially the most dangerous radiation if inside a body, but the least dangerous if outside.

Radioactive decay

- Radioactive decay is a <u>random process</u>. It is <u>impossible</u> to say exactly <u>when a nucleus will decay</u>. The process is totally <u>unaffected by conditions</u> such as temperature and pressure. It is possible, however, if there are enough nuclei present, <u>to predict how many of them will decay</u> over a period of time.

Half life

- The **amount of radiation emitted each second** (the **activity** of a source, A) depends upon **how many unstable nuclei are present and the type of nuclide**.
- **The activity of a source is measured in becquerels (Bq). A source which has an activity of 1 Bq is decaying at the rate of one nucleus per second.**
- As time goes by, the **number of unstable nuclei** in a sample **decreases**.
- The **activity of a source** therefore **decreases with time**.
- We describe this decrease using the idea of a **half life**.

The half life of a radioactive material is the average time for the number of undecayed nuclei in a sample of material to halve.

Different radioactive nuclei have different half lives. Although, as seen below, half lives can vary enormously, a graph of the **number of undecayed atoms** in a sample **against time** always has the **same shape**.

Isotope	Half life
Uranium–238	4500 million years
Radium–226	1620 years
Strontium–90	28 days
Radon–222	4 days
Radium–214	20 minutes

○ undecayed atom   ● decayed atom

large number of unstable nuclei – lots of radiation emitted

fewer unstable nuclei – less radiation emitted

half life

half life

after each half life the number of undecayed nuclei halves

after each half life the activity of the source halves

A graph of **count rate against time** has the **same shape** for all radioactive nuclei.

We can use these graphs to determine the half lives of radioactive nuclei.

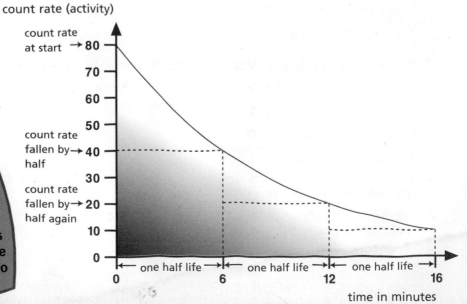

count rate (activity)

count rate at start → 80

70

60

50

count rate fallen by → 40 half

30

count rate fallen by → 20 half again

10

0

one half life   one half life   one half life

0   6   12   16

time in minutes

EXAMINER'S TOP TIP
Beware the common misconception that half a sample of radioactive materials decays in one half life therefore all the sample will decay in two half lives.

Half life calculations

Once we know the half life of a radioactive material, we can calculate the number of undecayed nuclei in a sample or its activity after a certain number of half lives.

EXAMPLE

The initial activity of a radioisotope is 960 counts per minute. If the half life of the isotope is 20 minutes, calculate the activity of the source after a) one hour b) one hour forty minutes.

$$960 \xrightarrow{\div 2} 480 \xrightarrow{\div 2} 240 \xrightarrow{\div 2} 120$$

after 1 half life — total time = 20 mins

after 2 half lives — total time = 40 mins

after 3 half lives — total time = 60 mins

$$30 \xleftarrow{\div 2} 60 \xleftarrow{\div 2}$$

after 5 half lives — total time = 100 mins

after 4 half lives — total time = 80 mins

The decay constant

The activity of a source is directly proportional to the number of undecayed nuclei present.

$$A \; \alpha \; N$$

or $A = \lambda N$ where λ is a constant known as the **decay constant** for that nuclei.

EXAMPLE

A pure sample of element X contains 3.0×10^{19} atoms. Calculate the activity of the sample if the decay constant for X is $0.015 \; s^{-1}$

$$A = \lambda N$$
$$A = 0.015 \times 3.0 \times 10^{19}$$
$$A = 4.5 \times 10^{17} \; Bq$$

The relationship between the decay constant and the half life ($t_{\frac{1}{2}}$) for a particular nuclide is described by this equation:

$$t_{\frac{1}{2}} = \frac{\ln 2}{\lambda} \quad \text{or} \quad t_{\frac{1}{2}} = \frac{0.693}{\lambda}$$

EXAMPLE

Calculate the half life of the element X in the previous example.

$$t_{1/2} = \frac{0.69}{\lambda}$$
$$t_{1/2} = \frac{0.69}{0.015}$$
$$t_{1/2} = 46 \; s$$

Quick test

1 Explain what is meant by the statement 'a sample of element Y has an activity of 10 Bq'.

2 What effect do these have on the activity of a source?

  a increasing its temperature

  b decreasing the pressure

  c adding a catalyst

3 A radioactive isotope has a half life of 10 minutes. Calculate the fraction of the isotope that remains undecayed after 1 hour.

4 Calculate the decay constant for a sample of element Y that has an activity of 5.0×10^{10} Bq and contains 3.0×10^{10} nuclei.

5 Calculate the half life of element Y in question 4.

1. On average, 10 nuclei of the element are decaying each second. 2. None will affect the activity of the source. 3. 1/64 4. 1.7 s⁻¹ 5. 0.41 s

85

Subatomic particles

Matter and antimatter

Most of the world around us is composed of three basic particles. These are the proton, the neutron and the electron. However, it is now understood that for each of these particles there also exists an <u>antiparticle</u>.

The antiparticle of an electron is called a <u>positron</u>, the antiparticle of a proton is called an <u>antiproton</u> and the antiparticle of a neutron is called an <u>antineutron</u>.

● An antiparticle has a mass equal to the particle but has opposite charge.

● Antiparticles may be produced during radioactive decay or by cosmic rays, but they don't exist for very long.

● If a particle and its antiparticle collide, they annihilate each other, resulting in a burst of energy in the form of photons.

Leptons

● Electrons and neutrinos are **fundamental particles**, i.e. they cannot be subdivided into anything smaller.
● They belong to a larger group of particles called the **lepton family**.
● There are two other members of this group. They are called **muons** and **tau** particles. Both of these particles have similar properties to those of electrons, apart from their masses. The mass of a muon is approximately 200 times that of an electron and the mass of a tau particle is approximately 3500 times that of an electron.
● There are antiparticles of the muon and the tau. These are called the antimuon and the antitau.
● Like the electron and the positron, there is a neutrino associated with the muon, the tau and their antiparticles.
● Neutrinos have no charge and a mass of almost zero.
● Each member of the lepton family has been assigned a number called a **lepton number**. In any reaction involving leptons, the lepton number before the reaction must equal the lepton number after the reaction, i.e. we use these numbers to check if a particular reaction is allowed.
● Protons and neutrons are *not* leptons and so are regarded as having a lepton number of 0.

This table summarises the members of the lepton family and some of their properties:

Particle	Symbol	Lepton number (L)	Charge compared with an electron	Mass compared with an electron
Electron	e^-	+1	−1	1
Electron-neutrino	ν_e	+1	0	0
Muon	$\mu-$	+1	−1	200
Muon-neutrino	ν_μ	+1	0	0
Tau	$\tau-$	+1	−1	3500
Tau-neutrino	ν_t	+1	0	0
Positron	e_+	−1	+1	1
Antineutrino	$\bar{\nu}_e$	−1	0	0
Antimuon	$\mu+$	−1	+1	200
Muon antineutrino	$\bar{\nu}_\mu$	−1	0	0
Antitau	$\tau+$	−1	+1	3500
Tau antineutrino	$\bar{\nu}_t$	−1	0	0

Using lepton numbers

The equations below show several possible reactions involving leptons. We can use lepton numbers to see if the reactions are allowed.

EXAMPLE

$p \rightarrow n + \beta^- + \nu_e$ Using lepton numbers, we have

$0 = 0 + (-1) + (+1)$

The lepton numbers balance so this reaction is allowed.

EXAMPLE

$p + e^- \rightarrow n + \bar{\nu}_e$ Using lepton numbers, we have

$0 + (+1) = 0 + (-1)$

The lepton numbers do not balance, so this reaction is not allowed.

Neutrinos and antineutrinos

When alpha particles are emitted from the nucleus of an atom they are **mono-energetic**, i.e. they all have the same kinetic energy. But when beta particles are emitted from the nucleus of an atom they have a range of energies.

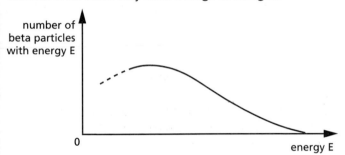

Observations of these energies, and the energies of the nuclides undergoing decay, seem to indicate that the law of conservation of energy is not being obeyed, i.e. it appears as if each nucleus must be losing energy in some other form. Observations of the decay of a nucleus also suggest that the law of conservation of momentum is also not being obeyed.

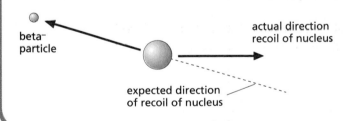

When a beta particle is emitted, the recoil velocity of the nucleus is not in an opposite direction to the beta particle. To solve these problems, it was suggested that another particle, as yet unknown, is being emitted by the nucleus during beta decay. This particle is called a **neutrino** and its antiparticle is called an **antineutrino**.

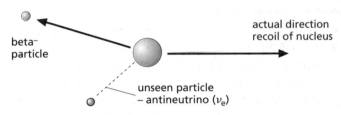

It is now accepted that during a beta-emission, an antineutrino ($\bar{\nu}_e$) is emitted and during a beta+ emission a neutrino (ν_e) is emitted.

$$^{40}_{19}K \rightarrow {}^{40}_{20}Ca + {}^{0}_{-1}\beta + {}^{0}_{0}\bar{\nu}_e$$
decay by emission of beta minus

$$^{11}_{6}C \rightarrow {}^{11}_{5}B + {}^{0}_{+1}\beta + {}^{0}_{0}\nu_e$$
decay by emission of beta plus

> **EXAMINER'S TOP TIP**
> Try this as an easy way to remember the groupings. There are three kinds of electron. There is an antiparticle for each of these. There is a neutrino associated with each of these. Making 12 leptons.

Quick test

1 What are the names of the antiparticles for
  a an electron **b** a neutrino?

2 What are the names of the three main groups of particles within the leptons?

3 Which two laws seemed to be disobeyed when a nucleus decayed by emitting a beta particle?

4 What is a fundamental particle? Give one example.

5 Describe what happens if a particle and its antiparticle collide.

6 Use lepton numbers to decide if this reaction is possible $n \rightarrow p + e^- + \bar{\nu}_e$

1. a) antineutrino b) positron 2. electrons, muons and tau particles (tauons) 3. law of conservation of energy and law of conservation of momentum 4. Fundamental particles are particles which cannot be subdivided into anything smaller. For example, leptons 5. They annihilate each other and there is a burst of energy. 6. $0 = 0 + (+1) + (-1)$ so reaction possible

Quarks and antiquarks

Forces within the nucleus

- There are strong electrostatic repulsive forces between the protons in a nucleus.
- There are also attractive gravitational forces between the nucleons in a nucleus, but these are very weak compared with the electrostatic forces.
- Therefore, for nuclei to be as stable as they are, there must be another attractive force strong enough to overcome the electrostatic repulsion and bind the nucleons together.

It is called the strong nuclear force.

- The strong nuclear force acts only over very small distances, for example, 1×10^{-15} m, i.e. it acts only between nucleons that are next to each other.

There is also a weak nuclear force

- The weak nuclear force acts over even shorter distances than the strong nuclear force.

Probing the nucleus

In the early 1900s scientists had discovered the 'nuclear nature' of the atom using the **alpha scattering experiment**. To determine whether the protons and neutrons in the nucleus were fundamental particles, or whether they were themselves composed of even smaller particles, scientists devised further scattering experiments. To penetrate the nucleus, **particles** with very **high energies (speeds)** were needed. Machines called **linear accelerators**, **cyclotrons** and **synchrotrons** were used to accelerate particles such as protons and electrons to very high speeds before being directed toward their target nuclei. After their interaction with the **nuclei** their **behaviour** and that of any particles that were created were observed. From these experiments, a very large number of **subatomic particles** have been identified, together with some of their **basic properties**, for example **mass**, **charge** etc.

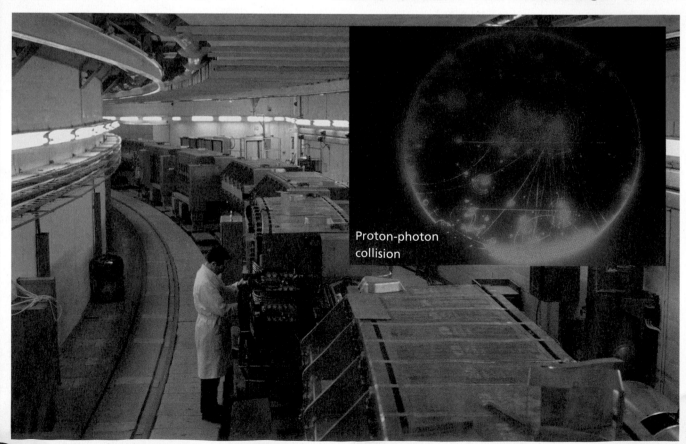

Proton-photon collision

The Proton Synchrotron particle accelerator at CERN, the European particle physics laboratory outside Geneva.

Quarks

It is now accepted that **leptons** are **fundamental particles**, but **protons** and **neutrons** are not. Protons and neutrons are composed of even smaller particles called **quarks**. There are **six** different **types of quark**, but only two of these are found in protons and neutrons. These are the **up quark** and the **down quark**. The up quark has the symbol **u** and has a charge of **+2/3 e**. The down quark has the symbol **d** and has a charge of **−1/3 e**.

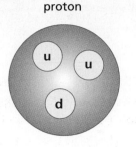
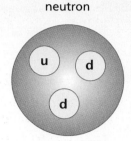

The quarks contained in nucleons

It follows that a proton therefore contains two up quarks and one down quark and a neutron contains one up quark and two down quarks.

- Particles that are made up of **three quarks** are called **baryons**.
- There are other particles besides the proton and neutron that contain different combinations of quarks but they have very short lifetimes and need not be discussed here.

Some particles are composed of a quark and an **antiquark**. These are called **mesons** and play an important role in nuclear bonding.

- Like leptons, baryons have a number assigned to them that allows us to see if a particular reaction is possible.
- The baryon number of an up quark or a down quark is +1/3.
- The baryon number of an antiquark is −1/3.
- Since a meson consists of a quark and an antiquark, its baryon number is 0.
- The mesons and baryons together form a large family of particles called **hadrons**.
- Hadrons are affected by both the **strong nuclear force** and the **weak nuclear force**.
- **Leptons** are only affected by the **weak nuclear force**.

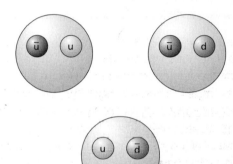

Examples of mesons

Quick test

1 What is a hadron?

2 What is a baryon?

3 What is the structure of these? **a** a proton **b** a neutron

4 Why are high-energy particles needed to explore the structure of the nucleus of an atom?

5 What is the structure of a meson?

6 What is the baryon number of a meson?

1. particles made up of quarks, i.e. baryons and mesons 2. particle made up of three quarks 3. a) two up quarks and one down quark b) two down quarks and one up quark 4. to penetrate the nucleus 5. one quark and one antiquark 6. 0

Exam-style questions
Use the questions to test your progress. Check your answers on pages 92–95.

1 The diagram shows the scattering of alpha particles as they pass through a very thin piece of gold foil.

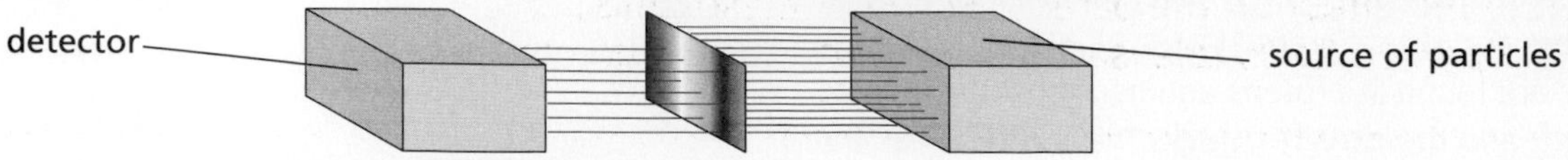

detector ———— source of particles

a Describe quantitatively what happens to the particles. [3]

....
....
....

b Describe what conclusions can be drawn about the structure of the gold atoms from this experiment. [3]

....
....

Look carefully at the paths followed by the three alpha particles drawn here as they pass through the nucleus of a gold atom.

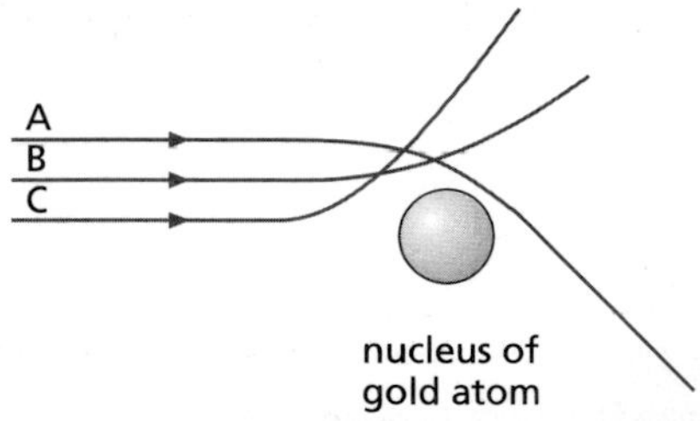

A
B
C

nucleus of
gold atom

c Which of the three paths A, B or C is not possible?
Give a reason for your answer. [2]

....
....
....

2 **a** Explain what is meant by the 'activity of a source'.
In what units would you measure activity? [2]

....

b You are given 1.0 g of pure $^{238}_{92}$U sample. If one nucleus of uranium has a mass of 3.95×10^{-25} kg, calculate the number of atoms in the sample. [2]

.... $1 \times 10^{-3} / 3.95 \times 10^{-25} = 2.532 \times 10^{21}$ ....

c Calculate the activity of the source if the half life of uranium–238 is 4.5×10^9 years. [2]

.... $A = \lambda N \qquad = 2.532 \times 10^{21} \times 4.5 \times 10^9$ ....
$= 1.139 \times 10^{31}$

3 **a** You are provided with a radioisotope that may be emitting alpha, beta or gamma radiations, or any combination of the three. Describe briefly an experiment you could carry out to determine which of the three types of radiation is/are being emitted. [5]

....
....
....

b Explain how it is possible for a nucleus to emit
i) an electron   **ii)** a positron. [2]

....
....

c The isotope $^{11}_{6}C$ decays by emitting a positron. Write an equation that describes this nuclear reaction. [2]

....

....

d Describe the effect on the parent nuclide of these emissions:
i) two alpha particles released in sucession   **ii)** a gamma ray. [3]

....

....

....

4 a Sketch a graph to show the range and distribution of the energies of beta particles emitted from a radioisotope. [2]

b Explain why the fact that beta particles have different energies when they are emitted suggests that other, unseen particles, are also being emitted. [2]

....

....

c Draw a diagram to show how the law of conservation of momentum, when applied to beta emission, also suggests that another particle is being emitted. [2]

d What is the name of this unseen particle? [1]

....

5 a What is the antimatter equivalent of an electron? [1]

....

b Describe what happens when a particle and its antiparticle collide. [2]

....

....

c Name three fundamental particles. [3]

....

....

....

d Use the conservation of lepton numbers to determine whether the reactions shown below are possible.
i) $p + \bar{\nu}_e \rightarrow n + \beta^+$ [2]

....

....

ii) $n \rightarrow p + \beta^- + \nu_e$ [2]

....

....

Total   /38

Answers

Forces and motion 1 (pp. 4–15)

1

$R^2 = 20^2 + 4^2 = 416$ $R = 20.4$ km h^{-1}

$\theta = \tan^{-1}\dfrac{4}{20} = 11.3°$

R is 20.4 km h^{-1} 11.3° west of north

2 a) a scalar has magnitude but no direction, for example mass, volume, energy, power, charge. A vector has magnitude and direction, for example displacement, velocity, acceleration, force, momentum

b) Tension in each rope in direction of motion) = $100 \times \cos 15 = 97$ N
Total force in the direction of movement = $2 \times 100 \cos 15 = 194$ N

c) Use $F = ma$ $a = \dfrac{F}{m} = \dfrac{194}{2000} = 0.097$ m s^{-2}

d) Water exerts drag on barge which balances forward forces, so once it is moving, no resultant force, so constant speed

3 a) $u = 45$ m s^{-1}
$t = ?$ $a = -3.0$ m s^{-2}
$v = 0$ m s^{-1}
$v = u + at$
$t = \dfrac{v - u}{a} = \dfrac{0-45}{-3} = 15$ s

b) $v^2 = u^2 + 2as$
$\dfrac{v^2 - u^2}{2a} = s = \dfrac{2025}{6} = 337.5$ m

c) $F = ma = 100\,000 \times 3 = 300\,000$ N (300 kN)

4 a) displacement = $20 \times 2 = 40$ cm

b) X at maximum gradient, i.e. displacement = 0

c) velocity approximately 1.2 m s^{-1}

5 a) $s = ut + \frac{1}{2}at^2$
$s = 300$ m $u = 0$ $t = ?$ $a = 9.8$ m s^{-2}
$300 = \frac{1}{2} \times 9.8 \times t^2$ $t^2 = \dfrac{600}{9.8}$ $t = 7.8$ s

b) $7.8 \times 30 = 234$ m

c) 30 m s^{-1}, 77 m s^{-1}

6 a) $v^2 = u^2 + 2as$
$0 = 625 + 2 \times -9.8 \times s$
$s = \dfrac{625}{2 \times 9.81} = 32$ m

b) $v = u + at$
$t = \dfrac{v - u}{a}$
$= \dfrac{0 - 25}{-9.8}$
$= 2.55$ s
total time ball is in the air = $2 \times 2.55 = 5.1$ s

c) $v^2 = u^2 + 2as$
$0 = u^2 + 2 \times 9.8 \times 64$
$u = 35$ m s^{-1}

7 a)   **b)**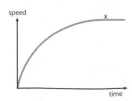

b) increase time of deceleration by flexing legs, reduce force since
$F = \dfrac{mv - mu}{t}$

Forces and motion 2 (pp.18–27)

1 a) Let velocity of two cars after collision be v.
Using conservation of momentum,
$1000 \times 25 = 2500 \times v$
$v = 10$ m s^{-1}

b) If elastic collision, E_k before collision = E_k after collision
E_k before = $\frac{1}{2} \times 1000 \times 25^2 = 312\,500$ J
E_k after = $\frac{1}{2} \times 2500 \times 10^2 = 125\,000$ J
so inelastic collision

2 a) Boat is stationary before man jumps off. Let velocity of boat after man jumps off be v.
Using conservation of momentum,
$0 = 80 \times 3 + 120 \times v$
$-240 = 120 \times v$
$v = 2$ m s^{-1} westwards

b)

exhaust gases gain momentum in this direction

rocket gains momentum in this direction

conservation of momentum and rocket propulsion

• Rocket propulsion uses principle of conservation of momentum
• Exhaust gases gain momentum in one direction, rocket gains momentum at same rate but in opposite direction

3 a) before impact: 10 m s^{-1}
after impact: −10 m s^{-1}
change in velocity = 20 m s^{-1}

b) change in momentum = $0.2 - -2.0 = 4.0$ kgm^{-1}

c) $F = \dfrac{mv - mu}{t} = \dfrac{4}{0.05} = 80$ N

4 since object in equilibrium:
sum of clockwise moments about X = sum of anticlockwise moments about X
$F = 0.05 = 4.0 \times 9.8 \times 0.40 + 1.5 \times 9.8 \times 0.20$
$F = 370$ N

5 a) $E_k = \frac{1}{2}mv^2 = 0.5 \times 0.4 \times 40^2 = 320$ J

b) Assuming energy conserved, E_p before release = 320 J

c) 320 J $= mgh = 0.4 \times 9.8 \times h$
$h = 82$ m

6 a) $E_p = mgh = 70 \times 9.8 \times 500 = 34\,3000$ or 340 kJ

b) power = $\dfrac{\text{work done}}{\text{time taken}} = \dfrac{343\,000}{5 \times 60} = 1143$ W or 1.1 kW

c) Some of the energy he obtains from his food is changed into heat energy rather than gravitational potential energy

7 a) power = $\dfrac{\text{work done}}{\text{time}} = \dfrac{240 \times 9.8 \times 4}{60} = 160$ W

b) power = force $\times$ velocity = $5000 \times 60 = 300\,000$ W = 300 kW

8 a) second hand makes one complete revolution in 60 s
$\omega = \dfrac{2\pi}{60} = 0.10$ rad s^{-1} (6.0° s^{-1})

b) $v = r\omega = 0.015 \times 0.10 = 0.0015$ m s^{-1}

Waves (pp. 30–41)

1 a) transverse: vibrations perpendicular to direction of travel; example: any electromagnetic wave
longitudinal; vibrations along, or parallel to, direction of travel, example: sound

b) $f = \dfrac{1}{T} = \dfrac{1}{2 \times 10^{-2}} = 50$ Hz

c) $v = f\lambda$ $\lambda = \dfrac{340}{200} = 1.7$ m

2 a) i)
ii)
iii)

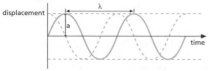

b) Use $I = \dfrac{P}{4\pi r^2}$
$P = 0.1 \times 100$ W $= 10$ W, $r = 2.0$ m
$I = \dfrac{10}{(4 \times 3.14 \times 4.0)} = 0.20$ Wm^{-2}

3 a) speed of light changes as crosses boundary, therefore since ray not striking boundary at 90°, light is refracted

b) i) $1.3 = \dfrac{\text{speed light in vacuum}}{\text{speed in water}}$
speed in water $= \dfrac{\text{speed in vacuum}}{1.3} = \dfrac{3 \times 10^8}{1.3} = 2.3 \times 10^8$ m s^{-1}

ii) $\dfrac{\sin 45}{\sin 38} = 1.15$

iii) $_wn_g = \dfrac{\text{speed of light in water}}{\text{speed of light in glass}} = 1.15 = \dfrac{2.3 \times 10^8}{\text{speed of light in glass}}$
speed of light in glass $= 2.0 \times 10^8$ m s^{-1}

4 a) $_1n_2 = \dfrac{2.0 \times 10^8}{1.8 \times 10^8} = 1.11$

b) $\sin C = \dfrac{1}{_1n_2} = 0.90$
$C = 64°$

c) optical fibres are: capable of carrying more information, free from noise, little loss in energy so can travel further without boosting, lighter and smaller, more difficult to tap conversations carried by fibres

5 a) sources of waves are coherent if: they have same frequency and they maintain a constant phase relationship

b) i) loudspeakers each act as source of sound, but sources coherent so they produce interference pattern with maxima and minima
ii) as he moves away from source, intensity will decrease so 'loudness' will decrease

6 a) i) $\sin \theta = \dfrac{\lambda}{a} = \dfrac{600 \times 10^{-9}}{0.1 \times 10^{-3}} = 0.006$
$\theta = 0.34°$

ii) $\dfrac{\text{distance } x}{1.0} = \tan 0.34$
$x = \tan 0.34 = 0.006$ m $= 6$ mm

b) white central fringe, edges tinged with red

Electricity (pp. 44–55)

1 a) lots of free charge carriers, electrons that are not held strongly by positive nuclei of metal atoms and are therefore able to drift from atom to atom

b) some liquids contain charge carriers called ions – when a potential difference applied between two electrodes, positive ions drift towards negative electrode and negative ions drift towards positive electrode

c) $I = nAve$
$v = \dfrac{I}{AAe} = \dfrac{1}{8.5 \times 10^{28} \times 2 \times 10^{-6} \times 1.6 \times 10^{-19}} = 3.7 \times 10^{-5}$ ms^{-1}

d) if wire warmed, vibrations of positive ions would increase, impeding movement of electrons, so average drift velocity would decrease

2 a) resistance is property of individual component, resistivity is property of material from which component made

b) $R = \dfrac{\rho l}{A}$
$R = \dfrac{1.55 \times 10^{-8} \times 0.75}{\pi \times (10 \times 10^{-3})^2} = \dfrac{1.1625 \times 10^{-8}}{3.14 \times 10^{-6}} = 0.004$ Ω

c) $R = \dfrac{\rho l}{A}$
$\dfrac{RA}{\rho} = l = \dfrac{30 \times \pi \times (0.025 \times 10^{-3})^2}{1.1 \times 10^{-6}} = \dfrac{5.8 \times 10^{-8}}{1.1 \times 10^{-6}} = 5.4$ cm

3 a) $P = IV$
$I = \dfrac{1.0}{5.4} = 0.19$ A

b) $I = 0.19$ A $V = 0.6$V $r = \dfrac{0.6}{0.19} = 3.2$ Ω
so each cell has internal resistance 0.8 Ω

c) $W = RI^2t = 0.8 \times 0.19 \times 0.19 \times 5 \times 60 = 8.7$ J

4 a) $\dfrac{1}{R_t} = \dfrac{1}{4} + \dfrac{1}{12} + \dfrac{1}{6} = 0.5$ $R_t = 2$ Ω

b) $\dfrac{1}{R_t} = \dfrac{1}{2} + \dfrac{1}{2} = 1$ $R_t = 1$ Ω

c) current I_t through circuit $= \dfrac{12}{3} = 4$ A
so current through each 2 Ω resistor $= 2$ A

d) pd across 4 Ω resistor $= \tfrac{2}{3} \times 12 = 8$ V ∴ $I = 2$ A

5 a) i) Let p.d. across thermistor be V_{th}
$V_{th} = \dfrac{120}{120 + 40} \times 6 = 4.5$ V

ii) $V_R = \dfrac{40}{120 + 40} \times 6 = 1.5$ V

b) If temperature of thermistor increases, resistance of thermistor decreases, so V_R increases and so V_{th} decreases.

c) temperature control circuit

6 a) Kirchhoff's first law: the algebraic sum of the currents flowing into a circuit junction must be equal to the algebraic sum of the currents flowing out of that junction
Kirchhoff's second law: the algebraic sum of the e.m.fs in any closed loop is equal to the algebraic sum of the p.ds around that loop

b) Using Kirchhoff's second law:
2 V $= 10I_1 - 10I_3$ [1]
4 V $= 10I_2 + 10I_3$ [2]
Using Kirchhoff's first law:
$I_1 + I_3 = I_2$ [3]
Substitute [3] into [2]
4 V $= 10(I_1 + I_3) + 10I_3$
4 V $= 20I_3 + 10I_1$ [4]
[4] – [1]: 2 V $= 30I_3$ $I_3 = \tfrac{2}{30}$ A $= \tfrac{1}{15}$ A
Substituting in [1]
2 V $= 10I_1 - \tfrac{10}{15}$
$\tfrac{8}{3} = 10I_1$ $I_1 = \tfrac{4}{15}$ A
Substituting in [3]
$I_3 + I_1 = \tfrac{1}{15} + \tfrac{4}{15} = \tfrac{1}{3}$ A

7 a) ohmic conductor obeys Ohm's law, example: metallic wire

b)

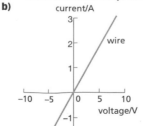

c) non-ohmic conductor: filament lamp

d)

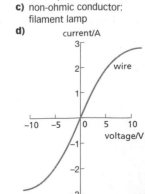

Quantum physics (pp. 58–63)

1 **a)** radio, microwaves, infra-red, visible, ultraviolet, X-rays, gamma rays

b) all able to travel through vacuum, travel at 3×10^8 m s^{-1} through vacuum, all transverse waves, all transfer energy, all can be reflected, refracted, diffracted and interfere

c) pass light through polariser, observe polarised light through analyser, turn analyser through 90°, if light transverse wave will not be able to pass through crossed polariser and analyser

2 **a)** Energy levels are unique to a given element, therefore the 'jumps' between the levels which produce the lines of the spectrum are unique to the element. Each element has distinct emission line spectrum that acts like a 'fingerprint' for that element.

b) As white light passes through gas surrounding star, the atoms of the gas may absorb some of the energy in order to make transitions to higher energy levels. The wavelengths of the absorbed light will exactly match the energy needed to make upwards transition. Spectrum of white light will show dark lines where these wavelengths are missing. These dark lines are characteristic of the atoms in the gas, therefore the atoms in the gas can be identified.

3 **a)** a small packet of light energy or a quantum of radiation

b) $E = hf$ $f = \dfrac{c}{\lambda}$

$E = 6.6 \times 10^{-34} = \dfrac{3 \times 10^8}{6.0 \times 10^{-2}} = 3.3 \times 10^{-24}$ J $= 2.1 \times 10^{-5}$ eV

$E = 6.6 \times 10^{-34} = \dfrac{3 \times 10^8}{5.0 \times 10^{-10}} = 3.96 \times 10^{-16}$ J $= 2500$ eV

4 **a)** A zinc plate is placed on cap of gold-leaf electroscope. Cap and plate are charged negatively. Ultraviolet radiation is shone on plate, gold leaf falls which shows electrons are being emitted.

b) 3.6 eV is minimum energy required to enable electrons to be emitted from surface of zinc.

c) $\Phi = hf_0$

$f_0 = \dfrac{2.7 \times 1.6 \times 10^{-19}}{6.6 \times 10^{34}} = 6.5 \times 10^{14}$ Hz

5 **a)**

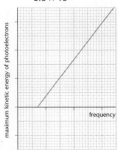

b) $\Phi = hf_0$

$hf = \Phi + \frac{1}{2}mv^2$

$hf - \Phi = \frac{1}{2}mv_{max}^2$

$6.6 \times 10^{-34}(6 \times 10^{14} - 4.0 \times 10^{14}) = \frac{1}{2}mv_{max}^2$

$= 1.32 \times 10^{-19}$ J $= 0.825$ eV

c) if photon interacts with electron within metal, some kinetic energy may be lost as the electron moves towards surface through collision, so photoelectrons emitted with range of energies

6 **a)** light can act as a wave (as in reflection, refraction, diffraction) and also like a particle as in photoelectric emission

b) $\lambda = \dfrac{h}{p} = \dfrac{6.6 \times 10^{-34}}{9.1 \times 10^{-31} \times 1.8 \times 10^7} = 4.0 \times 10^{-11}$ m

7 **a)** **i)** electron has lowest possible energy for that atom
ii) electron has gained energy and has moved to higher energy level

b) 1.9 eV

c) 13.6 − 1.5 = 12.1 eV

d) $E = hf = \dfrac{hc}{\lambda}$

$\lambda = \dfrac{hc}{E} = \dfrac{6.6 \times 10^{-34} \times 3 \times 10^8}{12.1 \times 1.6 \times 10^{-19}}$

$= \dfrac{1.98 \times 10^{-25} \times 10^{-18}}{1.94} = 1.0 \times 10^{-7}$ m

Kinetic theory (pp. 66–77)

1 **a)** to raise temperature of 1 kg of water by 1 K requires 4200 J of energy

b) heat liquid electrically in insulated container: note p.d. across heater, current flowing and time heater turned on, mass of substance and temperature rise caused by heating, use

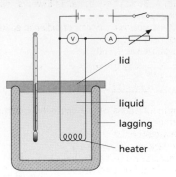

$IVt = mc\Delta\theta$ to calculate θ

c) to change ice to water no increase in temperature:
$\Delta Q = mL = 2.0 \times 300$ kJ $= 600$ kJ
to raise temperature of water to 100 °C: $\Delta Q = mc\theta = 2.0 \times 4200 \times 100 = 840\,000$ J
to change 2 kg water to 2 kg steam no change temperature:
$\Delta Q = mL = 2 \times 2260$ kJ $= 4520$ kJ
total needed $= 600$ kJ $+ 840$ kJ $+ 4520$ kJ $= 5960$ kJ

2 **a)** increasing temperature leads to atoms moving faster, more collisions each second and so increase in pressure

b)

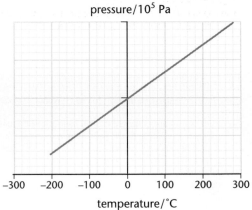

c) $\dfrac{V_1}{T_1} = \dfrac{V_2}{T_2}$

$V_2 = 2 \times 10^{-4} \times \dfrac{800}{400} = 4.0 \times 10^{-4}$ m^3

d) $PV = nRT$

$\dfrac{1 \times 10^5 \times 2.0 \times 10^{-4}}{8.3 \times 400} = n = 0.006$ moles

3 **a)** P is pressure of gas in Pascals, V is volume of gas in m^3, N is number of particles in gas, m is mass of each particle and $\overline{c^2}$ is mean of the squares of the speeds of the particles

b) atoms continuous and random motion, total volume of gas atoms negligible compared with volume of gas itself, molecules undergo perfectly elastic collisions, no intermolecular forces except during collisions

c) $\overline{c^2} = \dfrac{300^2 + 350^2 + 400^2 + 450^2}{4} = \dfrac{575\,000}{4} = 143\,750$ m^2 s^{-2}

$\sqrt{\overline{c^2}} = 379$ m s^{-1}

d) $P = \frac{1}{3}\rho\overline{c^2} = \frac{1}{3} \times 1.5 \times 550^2 = 1.5 \times 10^5$ Pa

4 **a)** Applied force and cross-sectional area of wire − F and A

b) Original length of wire and increase in length of wire due to load − Δl and l

c) $E = \dfrac{\text{stress}}{\text{strain}} = \dfrac{3.0 \times 10^8}{2.0 \times 10^{-3}} = 1.5 \times 10^{11}$ Pa

d) energy $= \frac{1}{2} \times$ stress $\times$ strain $= 0.5 \times 3.0 \times 10^8 \times 2.0 \times 10^{-3}$
$= 3.0 \times 10^5$ Jm^{-3}

5 **a)** force initially proportional to extension, reaches limit of proportionality, then elastic limit, then yield point and finally breaks

b) copper

6 **a)** ΔU is change in internal energy of system
ΔQ is energy entering system by heating
ΔW is energy entering system by work being done on it

b) heat engines convert heat into work, heat pumps move heat from places of lower temperature to places of higher temperature

c) Maximum efficiency $= \dfrac{T_1 - T_2}{T_1} = \dfrac{500 - 300}{500} = 0.4$ (or 40%)

The nuclear atom and radioactivity (pp. 80–89)

1 a) almost all pass through foil, a small number are deflected, a very small number are scattered backwards

b) atoms mainly empty space, tiny regions of concentrated positive charge, which explains back-scattering as repulsion of alpha particles by similarly charged objects

c) In path A, particles are attracted towards nucleus, which would not happen.

2 a) activity of source is amount of radiation emitted each second, measured in becquerels (Bq)

b) mass of sample = 0.001 kg
$$\frac{1 \times 10^{-3}}{3.95 \times 10^{-25}} = 2.53 \times 10^{21}$$

c) $A = \lambda N = \dfrac{0.69 \times 2.53 \times 10^{21}}{4.5 \times 10^{9}} = 3.9 \times 10^{11} = 1.2 \times 10^{4}\ \text{s}^{-1}$

3 a) different radiations will penetrate different thicknesses of material
alpha radiation can only travel short distances in air, so will not reach detector if distance is more than few centimetres
beta radiation will travel up to metre in air so will be detected if distance less than 1 m
gamma radiation will travel almost unlimited distances so will be detected if distance greater than 1 m

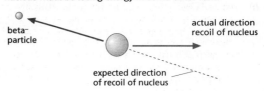

b) i) β– particle created in nucleus when neutron changes into proton, electron and an antineutrino – proton remains in nucleus, electron and an antineutrino are emitted
ii) β+particle created when proton in nucleus changes into neutron and emits a positron

c) $^{11}_{6}\text{C} \longrightarrow {}^{11}_{5}\text{B} + {}^{0}_{+1}\text{e}$

d) i) nucleon number decrease by 8, proton number decrease by 4
ii) proton and nucleon number unchanged

4 a)

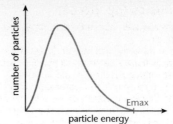

b) observation of this range of energies suggests that law of conservation of energy not being obeyed, it appears that each nucleus must be losing energy in some other form

c)

beta⁻ particle — actual direction recoil of nucleus — expected direction of recoil of nucleus

d) neutrino

5 a) positron
b) they annihilate each other and release ray photons
c) electrons, neutrinos, muons, tau, quarks
d) i) $0 + -1 \longrightarrow 0 + -1$ so reaction possible
ii) $0 \longrightarrow 0 + 1 + 1$ so reaction not possible

Index